The ENERGY GUIDE

This book is dedicated to the farmers who care for their soil and grow our food, allowing us to cultivate great energy. Thank you.

DR LIBBY WEAVER

The ENERGY GUIDE

A STEP-BY-STEP PLAN TO FINDING THE ENERGY YOU NEED TO FLOURISH

Pan Macmillan Australia

CONTENTS

Parts of *The Energy Guide* have been adapted from Dr Libby Weaver's *Exhausted to Energized*, released 2015. All recipes are original.

INTRODUCTION

For too long, people in the Western world have used weight as a benchmark for health. Yet both science and clinical experience demonstrate that when we focus on weight, we are more likely to regain anything we lose, plus more. A far more powerful insight into our level of wellbeing is our energy level, so much so that I want it to be seen as the true health currency.

I want people to be curious about why they are tired, and passionate about resolving it. Because, let's face it, everything is more difficult when we are exhausted.

Every day in my practice I see otherwise healthy people in their twenties, thirties, forties and fifties telling me that they feel exhausted and overwhelmed. When their alarm goes off in the morning, they don't bound out of bed, grateful to be alive, excited to discover what lies ahead. They wake up feeling tired, dreading how they are going to get through the day.

It doesn't need to be this way. It's time people understood what levers to pull to get their energy back. It's time to make your energy your highest priority for, without it, you are only half alive.

The first step is to listen to your body – it will tell you when it's time to slow down, rest, better support and enhance detoxification pathways, repair, replenish and restore. Yet too many people ignore this, throw down another pill, or write it off as 'I must be getting old'.

(As an aside, I have had 25-year-olds tell me they are really tired, believing it must be because they are getting old. And my reply is, 'Hashtag, you're joking!' If you are 25 and exhausted, you need to do something NOW to get to the heart of why that is. Youthful energy is, by its very definition, strong, consistent and boundless.)

Humans have been on this planet for about 150,000 years and, up until the very recent past, have lived and worked in harmony with the seasons – we had no other choice. From this way of living many interior bodily rhythms were created, one of which is our circadian rhythm. This rhythm contributes to the governance of many systems and processes inside the body, including the release of hormones, digestive enzymes, neurotransmitters, body temperature, sleep–wake cycles and other important bodily functions. Abnormal circadian rhythms have been linked to various sleep disorders including insomnia, as well as obesity, type 2 diabetes, depressed mood, bipolar disorder and seasonal affective disorder.

When we consider how our ancestors lived – before marketing started instructing our way of life – we see that our bodies are not built to be sedentary (or to run marathons, either!). We are not designed to live without adequate sunshine or be separate from Nature and her rhythms. We are not made to exist on very little sleep, or eat 'numbers' pretending to be food. We are not built to subsist on no-fat or no-carb diets. Nor are our brains wired to handle consistent and profound amounts of mental and emotional stress. We are also not designed to live without fresh air, yet I regularly meet people who wake up in an air-conditioned apartment, catch an elevator to their basement car park, get into an air-conditioned car and drive to work where they park underneath the air-conditioned office in which they spend their day, before repeating the action in reverse.

Our bodies are geared to survival and will choose any number of ways to let us know we are physically, mentally and/or emotionally exhausted: fatigue, low mood, anxious feelings, apathy, unrefreshed sleep or insomnia, brain fog, lowered resistance to infections, stiffness, digestive problems, 'unexplained' changes in body fat levels and signs of rapid ageing.

> 'Energy is the true health currency.'

When people experience low energy levels, they might try to sleep more, and for some this works. For others, though, it doesn't pay off and they wake up just as tired as when they went to bed. Of course, improving the quality of your sleep *will* help if you know that this is poor, but it won't resolve your personal energy deficit if your tiredness is the result of dysfunction in another body system caused by a nutrient deficiency, toxin overload or the stress of unresolved emotional issues.

As we will see, poor energy is often the result of the interplay between your biochemistry, nutritional status and emotional landscape. I call these my 'three pillars' of health and wellbeing. In Part One of this book, I help you explore how each of these pillars influences your experience of vitality and energy.

In Chapter One, I talk about how **what we eat** affects our energy levels, in particular how eating real wholefoods provides our body with the energy we then get to experience. I explain that real food is what the digestive system was designed to deal with, and also why good hydration is so crucial to feeling alive and alert.

Chapter Two is all about the importance of **sleep**, something so many people struggle with these days. You'll see many of the factors in modern life that can interfere with waking up refreshed, and simple changes you can make to help you sleep more restoratively.

In Chapter Three, I look at the role of **movement** in promoting energy, from building muscle mass to ensuring we stay flexible to the importance of sitting less.

Chapter Four is all about **stress**, and how our body's natural response to 'perceived danger' influences *every* aspect of our wellbeing, including how effectively we use body fat as a fuel, our sex-hormone balance and which foods we crave. I also share the most effective, science-based strategies for managing stress.

In the final chapter, I talk about some of the specific **health challenges** that affect energy, such as viral infections, allergies, liver congestion and thyroid dysfunction. I also talk about postnatal depletion – a debilitating experience for too many mothers these days – and the critical importance of women ensuring their health and energy reserves are replenished.

Part Two of this book features over 100 tasty and nutritious recipes, using real, energy-giving wholefoods that will nourish and sustain you.

As you read this book, you may begin to realise what steps you need to take to live in a new way, with a new level of self-care that restores you. For some, this will mean making changes to how you eat; for others, it will be about how you manage your time. But whatever you do, please do something. Continuing to do what you have always done and expecting a different result is some people's definition of insanity! Something has to change – the way you eat, drink, move, sleep, breathe, think, believe or perceive.

The good news is that making positive change in just *one* of these areas leads to an upward spiral in the others. Yet, some people won't do this for themselves due to an underlying belief that they aren't worth looking after; that they aren't worthy of their own love and care. Years of working with people has shown me this, day after day.

So I encourage you to explore this within yourself. What is stopping you taking one step towards eating, moving or sleeping in a way that better serves your energy? For these are the most basic foundations of wellness and we cannot fight the needs of our biology: we must get plenty of nutrients and few, if any, problematic substances from our food; we must move our bodies rather than sit for long periods; we must get seven to nine hours sleep per night; and we must embrace a daily restorative practice in our lives, even if it is just diaphragmatically breathing while we gaze out the window. If you can't take care of you for your own sake, do it for your family, your friends and your work. What is lousy energy costing you?

For every moment, we get to choose what we focus on. So the real question is: Do you feel the desire, the need or the longing to live in a different way? To live with ease and spaciousness and more energy? To stop punishing yourself or telling yourself that you are a failure and are not okay the way you are? The punishment has to stop, not only in your relationship with food and lifestyle choices, but in your relationships with other people, with work, with money and, most importantly, with yourself. It is your way of life that this book is about. My intention is to offer you new insights, and the renewed energy to live your best life; a life with a reservoir of abundant energy, allowing you to experience all that is on offer, and enabling you to share your gifts.

♡ Dr Libby

'What is stopping you taking one step towards eating, moving or sleeping in a way that better serves your energy?'

'Remind yourself of the difference between being alive and living. It matters.

And great energy helps you enjoy more of both.'

PART ONE

GETTING YOUR ENERGY BACK

CHAPTER ONE

ARE YOU EATING WELL?

The only way you obtain energy on a physical level is through how you eat. Why would you want to make choices that interfere with your experience of energy? Meeting your nutritional needs is essential for energy and vitality, allowing you to move, think, relate, work and play. The challenge for so many people, however, is that they do not feel energised by what they are eating, yet they often make no mental link to the consequences of their choices. You can cultivate exceptional energy in the way that you nourish yourself. In this chapter I show you how nutrients affect your energy levels and why nothing can replace a real-food, wholefood, nutrient-dense way of eating. Nothing.

Components of food can be categorised as macronutrients or micronutrients. The former are carbohydrates, proteins and fats (alcohol is also technically a macronutrient as it is a source of calories, but it's a challenge to put 'alcohol' and 'nutrient' in the same sentence – except for when it occasionally might nourish the soul!). We need these in large (macro) amounts for growth and metabolism, as fuel and for all body functions; they provide energy. The latter are vitamins and minerals, which we need in smaller (micro) amounts to enable us to extract the energy from the macronutrients we consume. As you will see, when you choose real wholefoods (minimally processed foods as close to their natural state as possible), you are well on the way to experiencing optimal energy levels.

THE MACRONUTRIENTS

Carbohydrates, proteins and fats provide energy, measured as calories or kilojoules (calories being the imperial and kilojoules being the metric). While each of these macronutrients provides energy, the amount varies.

1 g carbohydrate = 4 calories
1 g protein = 4 calories
1 g fat = 9 calories

However, not all calories are created equal. In other words, the number of calories does not take into account what I refer to as the 'metabolic consequences' of their consumption. For example, when you consume foods that contain carbohydrates, your pancreas secretes insulin to deal with the increase in blood glucose, and insulin signals to the body to store fat. So the metabolic consequence of calories consumed from carbohydrates is insulin production. Dietary fat, on the other hand, does not drive the production of any hormones that signal fat storage. Therefore, there aren't metabolic consequences of consuming calories from fats. You just get the energy. This is one reason that calorie-counting is an outdated weight-management strategy.

THE KREBS CYCLE

The biochemistry of energy is a big topic, and one that involves a host of magnificent pathways, all of which run on nutrients. The Krebs cycle, or citric acid cycle, is the central metabolic pathway in all aerobic (oxygen-utilising) organisms. Keeping it simple, it is a series of chemical reactions that occurs in the mitochondria of every cell in our body.

Picture a cell as a circle. Each of us is made up of about 50 trillion cells, a number so enormous it easily goes over our heads. Let me use a time analogy to demonstrate just how large that number is. One million seconds ago was 12 days ago. One billion seconds ago was 32 years ago. Yet one trillion seconds ago was 32,000 years ago!

Now imagine that inside almost all of these 50 trillion cells is a miniscule mouse on a treadmill generating energy. This energy-generating part of the cell is known as the mitochondria.

Mitochondria are cellular organelles that function as power plants within a cell – the mouse within the circle. They are responsible for the production of energy derived from the breakdown of carbohydrates and fatty acids. Mitochondria oxidise, or 'burn', the macronutrients for energy, yielding adenosine triphosphate (ATP), the cellular form of energy utilised for cellular processes all throughout the body. ATP provides the energy to pump your heart, power neurons in your brain, contract your muscles, exchange gases in your lungs, extract nutrients from food and regulate body temperature, amongst other things. Simply stated, mitochondria produce ATP, and ATP is the energy for your life – it is essential for survival. Without a sufficient generation of ATP, life would cease to exist.

CARBOHYDRATES

Carbohydrates supply the body with glucose, the preferred fuel for your brain, kidneys and red blood cells, and the fuel you use first when you exercise. Carbohydrates can be sweet or starch-based.

Sweet carbs include sugar, lollies, soft drinks and fruit juice. These are easily stored as body fat because you typically eat more than you can use for energy at that time. They usually offer little or no nutritional value. In fact, they may actually be detrimental to your health.

Starch-based carbohydrates include sweet potato, corn, pumpkin and brown rice. The energy from starch-based foods typically (although not always) takes longer to enter the bloodstream and provides a more sustained form of energy, often due to the fibre these foods contain.

Glycaemic index and load

The glycaemic index (GI) is a rating of how quickly the glucose from a food will hit the bloodstream. Essentially, the faster it gets to the bloodstream, the more insulin is required to mop up the excess glucose. However, the GI does not take into account the *amount* of carbohydrate in a food.

Glycaemic load (GL) is a better indicator of how a carbohydrate food will affect blood glucose levels. GL considers the actual amount of carbohydrate you eat in a sitting. For example, some types of carrots are considered a high GI carbohydrate, but are also high in fibre, water, vitamins, minerals and protective phytonutrients, which alters the way they are digested and promotes excellent health. As a result, their GL is low (a good thing).

Many chocolate bars, on the other hand, have a low GI, yet they contain high levels of refined sugars, poor-quality fats, virtually no nutrients and are high GL.

In other words, the GI can be highly misleading if that is all you use to guide your carbohydrate choices. No wonder people feel confused and overwhelmed at times about dietary information! However, we don't need science or labels to tell us that a carrot is better than a chocolate bar. Common sense is a better guide, or simply embracing the statement 'Just eat real food'.

Worth noting is that both GI and GL are affected by the protein, fat and fibre content of a meal, as these all slow down the release of glucose into the blood and hence the requirement for insulin (also a good thing). You don't want too much insulin being secreted or big surges of it. Elevated blood insulin damages the lining of blood vessels, signals body fat to be stored, interferes with the body's ability to manage appetite, stops the thyroid working properly and increases the risk of developing type 2 diabetes, all of which impacts energy, how your clothes fit, longevity and quality of life.

Fructose

Fructose has a low GI too. The way it is digested and utilised by the body is different from glucose. It requires an enzyme from the liver to convert it into glucose. Only liver cells can take up fructose, whereas all cells in the body can utilise glucose, and the impact of excessive fructose consumption on liver health is finally coming to light. A 'new' disease is now listed in medical textbooks called Non-alcoholic Fatty Liver Disease (see page 70). Up until the recent past, the only fat accumulation really ever seen was in the liver of chronic alcoholics, a condition known as Chronic Alcoholic Fatty Liver Disease. Now chronic fat accumulation in the liver is being found in people as young as teenagers, simply from the amount of processed foods and sugary drinks they've consumed. The consequences of this are major and debilitating and it is unacceptable that it is even happening.

Fibre

Fibre refers to certain types of carbohydrates that our body cannot digest. They can be soluble or insoluble. These carbohydrates pass through the intestinal tract intact and help to move waste out of the body. Diets that are low in fibre can cause problems such as constipation and haemorrhoids, and may increase the risk of certain types of cancers. Diets that contain more fibre have been shown to decrease the risk of colon cancer. The sources of fibre matter though.

For some people with specific bowel issues such as irritable bowel syndrome (IBS), high-fibre diets may work against optimal bowel health – and leave you feeling exhausted – unless you are careful to eat the right kind of fibre from food sources that suit your gut-health situation. Some people do well eating in a low FODMAP (fermentable oligosaccharides, disaccharides, monosaccharides and polyols) way, although clinically I've found that sometimes it is just certain foods within these fermentable groups that pose a problem.

Wholefoods high in fibre that generally don't tend to aggravate the bowel include vegetables, some starchy carbohydrates such as sweet potato, and some but not all fruits. In such cases, nutritional advice is best tailored to the individual.

PROTEIN

Protein supplies the body with amino acids, which are used to create the cells of the immune system. Amino acids also create neurotransmitters, which influence mood and create the muscles that provide us with the strength to carry our groceries and our children.

Of the 20 standard amino acids, 10 are considered essential (some research papers suggest nine are essential); in other words, you need to consume them in your diet as your body is unable to create them from other substances. A protein food that contains all of the essential amino acids is referred to as 'complete'.

Animal-based options, such as chicken, beef, lamb and fish, are all complete proteins. However, some people don't consume meat for a number of reasons, and many people would benefit from reducing their meat consumption. Remember that the properties and quality of the meat partly depends on the health of the animal in its lifetime, much of which is dependent on what it ate. If it consumed food that its species was designed to eat, the animal will have been healthier, required fewer, if any, antibiotics, and have a fat profile more supportive of human health than an animal raised on 'food' it was never intended to eat. So if you eat meat, always consider how the animal was raised and what it was fed.

Non-meat foods can also be good sources of protein (see over the page). Most plant-based protein sources are 'incomplete'; however, when you combine plant-based protein sources from different botanical families, you complete the protein, as the essential amino acid lacking in one of the foods will be high in the one from the other family. An example of this is adding sesame seeds to a vegetable stir-fry served with brown rice: the seeds and rice are from different families, and their amino acid profiles complete each other.

> 'When you choose real wholefoods, you are well on the way to optimal energy levels.'

Vegetarian sources of protein

NUTS & SEEDS	Nuts in general pack a mighty nutritional punch, providing protein and many essential vitamins, such as A and E, which are great for skin and heart health. They are also a good source of minerals (such as phosphorus and potassium), essential fatty acids and fibre. Raw nuts, seeds, or nut butter are a fantastic addition to any meal, as the protein and good fats keep you feeling fuelled for longer. Yet even though they contain nourishing properties, you can still over-consume them, mostly because many nuts are high in the pro-inflammatory omega-6 fats. In the not-so-distant past, we had to crack nuts or peel them before we could eat them, which slowed down our consumption. Nowadays it's all done for us, and it is all too easy to overeat them. A handful a day, though, is a great choice. Keep in mind that most of the problematic, excessive omega-6 fats come from processed foods, so if you don't eat many of these, obtaining some omega-6 fats from nuts is beneficial. (And to throw another spanner in the works, one of the omega-6 fats – linoleic acid – is essential, and we must eat it every day.) Good nut choices include macadamias and walnuts. Nuts range in protein content from 18 to 25 grams per 100 grams.
EGGS	Egg protein is of great biological value, meaning it is easily absorbed and utilised by the body. Egg yolks contain vitamin B12, which is of extra importance for vegetarians as animal foods are the only food sources of vitamin B12. Vitamin B12 improves your body's ability to use iron and generate red blood cells, which may help prevent anaemia, giving you more energy. Eggs make a fantastic breakfast option, especially when served with greens, and they offer 12 grams of protein per 100 grams.
QUINOA	Quinoa is another excellent source of protein, and is one of the only plant-based complete proteins. Remember, a 'complete' protein means that it contains all of the essential amino acids. Quinoa provides about 14 grams of protein per 100 grams. It is a wonderful option for people with coeliac disease or those with gluten sensitivity, as it is gluten-free. Botanically it is actually a seed, not a grain. As with all foods, though, let your body be your guide,and if you notice that quinoa upsets your digestion, park it.
CHICKPEAS	Chickpeas are a great source of fibre for vegetarians and contain 13 grams of protein per 100 grams. They are easy to incorporate into recipes as they don't have a strong flavour, and can be used with many different herbs and spices. If you have irritable bowel syndrome (IBS), they may not suit your digestive system until the IBS is resolved.

GREEN VEGETABLES	Perhaps a surprise inclusion in this list, leafy green vegetables contain amino acids, the building blocks of protein. When you eat large quantities of green leaves you supply your body with these amino acids. There is a growing body of evidence showing that so-called 'low-quality' plant protein actually allows for slow but steady synthesis of new proteins in the body. You only need to consider the muscle mass of gorillas for evidence of this. From a DNA perspective, gorillas are similar to humans, so the diet they naturally choose in the wild is of great interest to researchers, as their food choices haven't been influenced by marketing. Given that gorillas have one of the highest strength ratios of any creature and their diet is 55 per cent green leaves and around 5 per cent animal (e.g. insect) derived protein, it shows you the power of the amino acids in the greens.
LENTILS	Delicious, nutritious and super-easy to prepare, lentils can contain up to 18 grams of protein per 100 grams. Lentils also provide energy that is released slowly, fuelling you for longer. Nutty and earthy in flavour, they are easy to incorporate into any meal. Always soak and rinse them well before cooking.

ARE PROTEINS USED AS FUEL?

Protein-based foods supply the body with energy, although it is not directly used as fuel. Proteins are broken down into amino acids, and those amino acids can be converted into glucose to be used for energy, if needed. The name of the main biochemical pathway responsible for this is gluconeogenesis. Because biochemically it is more efficient for the body to use glucose supplied as glucose – it is more labour-intensive for the body to convert amino acids into glucose – the body prefers to allow the amino acids to be used for repair work, neurotransmitter production, muscle building and new cell creation, as no other substances can do this.

FATS

Fats from wholefood sources, which include avocados, nuts, seeds, oily fish and pasture-fed animals (I count butter in this category as well), provide us with energy. They also supply essential fatty acids that our bodies cannot make but must obtain from our daily food intake. Fatty acids play a crucial role in maintaining the health of your precious organs and also help the body to absorb the fat-soluble vitamins A, D, E and K, which help mediate inflammation. Fatty acids are also essential for the health of your skin, eyes, nails and hair. Consuming the right kind of fat will help you feel happy and content and fuller for longer, supporting sustained energy levels. Most people also find their desire for sweet food diminishes when they have more wholefood fats in their daily diet.

Clearly, there are immense health and energy benefits from eating the right fats. However, I have met thousands of clients who are afraid of consuming fat because of the bad press it has received in recent decades. Since the early 1980s people have been told that eating fat will make them fat. Yet for the majority

'Consuming the right kind of fat will help you feel content and fuller for longer, supporting sustained energy levels.'

of the population, cutting out or eating less fat hasn't made them any healthier or slimmer, for a multitude of reasons. One is that 'fat-free' products typically use refined sugars (or artificial sweeteners) to compensate for loss of flavour and texture and, to mask the increase in sweetness, a bucket-load of poor-quality salt is then added! Through the low-fat era, the Western world ate more sugar and salt than ever before in human history, in the false belief that this was a way to eat that supports health. This is another reason I recommend you don't get caught up in fads masking themselves as the 'right' way to eat, and simply eat real wholefoods.

Nourishing fats

There are three main categories of fats that serve our health: saturated, monounsaturated and polyunsaturated. All three can be obtained from eating wholefoods. And the fats that are detrimental to our health? Most concerning are the trans fats and poor-quality vegetable oils found mainly in processed foods, particularly deep-fried foods, cakes, biscuits and muesli bars – basically anything with a long shelf-life. Trans fats are created through a process called hydrogenation, where liquid plant oils are made into solids by the addition of hydrogen atoms. The name comes from the Latin word *trans*, meaning 'across', because the hydrogen atoms in the double bond are actually across from each other. Trans fats are known to interfere with many critical processes in the body, including cholesterol metabolism.

The essential fatty acids: omega-3 and omega-6

The omega-3 fats that must be consumed daily are eicosapentaenoic acid (EPA, found in flaxseeds, chia seeds and walnuts) and docosahexaenoic acid (DHA, found in cold-water fish such as salmon and mackerel). The body can make EPA and DHA from an omega-3 essential fatty acid called alpha-linolenic acid (ALA); however, the conversion is inefficient in some people. Hence, it is wise to include adequate amounts of all these fats in your diet.

The health benefits of EPA and DHA are well documented. These fats take up residence in the membrane (outer layer) of the cell and are able to exert their anti-inflammatory effects and keep the cells flexible. Due to their physical structure, however, they oxidise easily and are best consumed with antioxidant-rich foods (plenty of vegetables).

The omega-6 essential fatty acid is called linoleic acid (LA), which the body converts into the vital gamma linolenic acid (GLA) used, for example, in the production of sex hormones and skin health as well as immune function. This fat is found in evening primrose oil, borage oil and blackcurrant oil.

The ratio of fats consumed can be a problem, however. The average Western diet is overabundant in omega-6 fatty acids (typically from seed oils such as safflower oil, sunflower oil and soybean oil found in many processed foods) and lacking in omega-3s. This promotes widespread inflammation in the body, which is one of the ways we age from the inside out, and one of the mechanisms that creates degenerative diseases.

Fat metabolism and energy

All fats and oils are composed of fat molecules called fatty acids, which join together to make short-chain fatty acids (SCFAs), medium-chain fatty acids (MCFAs) or long-chain fatty acids (LCFAs). Three fatty acids joined together make a triglyceride, so you can have short-chain triglycerides (SCTs), medium-chain triglycerides (MCTs) or long-chain triglycerides (LCTs).

The majority of the fats and oils you eat – whether they are saturated or unsaturated, or come from an animal or a plant – are composed of long-chain triglycerides. In fact, studies suggest that 98 per cent of the fats we eat consist of LCTs.

LCTs require pancreatic enzymes to break them into smaller units. They are then absorbed into the intestinal wall and packaged into bundles of fat (lipid) and protein called lipoproteins. These lipoproteins are carried by the lymphatic system, bypassing the liver, and are dumped into the bloodstream, where they are circulated throughout the body. As they circulate in the blood, their fatty components are distributed to all the tissues of the body and they get smaller and smaller. When there is little left of them, they are picked up by the liver, broken apart, and used to produce energy or, if needed, repackaged into other lipoproteins and sent back into the bloodstream to be distributed throughout the body. Cholesterol, saturated fat, monounsaturated fat and polyunsaturated fat are all packaged together into lipoproteins and carried throughout the body in this way.

However, medium-chain triglycerides are not packaged into lipoproteins but go straight to the liver. Let me explain. MCTs, like those in coconut oil and organic butter, are broken down almost immediately by enzymes in the saliva and gastric juices. This means pancreatic fat-digesting enzymes are not essential for their digestion. In the digestive system, MCTs are broken down into individual fatty acids (MCFAs). Unlike other fatty acids, MCFAs are absorbed directly from the intestines into the portal vein (a vein that goes straight to the liver) and sent immediately to the liver where they are, for the most part, burned as fuel. They can therefore offer the body an energy supply that is highly efficient – provided your body gets the message from your nervous system that it is 'safe' for you to use fat as a fuel (see pages 49–50 for more on this).

One MCFA that deserves special mention is lauric acid (LA). Found in coconut and butter, lauric acid is known for its contribution to a healthy gut bacteria profile, essential for everything from an appropriate and efficient immune response to a good frame of mind. As an aside, butter from pasture-fed cows also contains conjugated linoleic acid (CLA), which research has shown boasts anti-cancer properties as well as anti-inflammatory actions.

THE MICRONUTRIENTS

Micronutrients – vitamins and minerals – literally keep us alive. Without them we die; they are that important. Vitamins and minerals themselves don't contain energy, but they allow the energy from the fats, proteins and carbohydrates we eat to be utilised.

B VITAMINS

Once we chew and swallow our food, it follows intricate biochemical pathways to become adenosine triphosphate (ATP), the substance the cells in our muscles, brain and other organs use for energy (see page 13). The most important group of nutrients in the conversion of food into ATP are the B vitamins.

There are many B vitamins, but three of them in particular are essential to this process: thiamin (vitamin B1), riboflavin (vitamin B2) and niacin (vitamin B3). Without these helpers, known as cofactors, energy conversion can be slow or disrupted, leaving us feeling sluggish and tired.

B vitamins are water-soluble, meaning that whatever is not used by the body is excreted. Therefore, B vitamins must be consumed every day. Fortunately, they are found in a wide range of foods.

- Thiamin-rich foods include lentils, nuts, seeds, tuna and pork (if you eat pork, make sure it is free-range and organic).
- Riboflavin is found in leafy green vegetables, tomatoes, almonds and eggs.
- Niacin is found in highest concentrations in animal protein such as beef, pork, chicken and fish. It is also found in beans.

Grains are also considered to be a good source of many B vitamins. If your digestive system can handle them, including high-quality grains such as spelt, oats and rye in your diet is a way to boost your vitamin B intake. Quinoa (not a true grain – it's botanically a seed) is gluten-free and also contains a variety of B vitamins.

If you are coeliac, gluten-intolerant or simply feel better without grains in your diet, it is important to ensure that you are getting adequate vitamin B from other foods. It may be worthwhile taking a good-quality multivitamin or a B-complex supplement to cover your vitamin needs, something best guided by an experienced health professional.

It is important to note that B vitamins are destroyed by heat and light, so the longer it takes a food to arrive on your plate after harvesting, the lower the levels. Also, research shows that when foods containing vitamins B1, B2 or B3 are heated over 90°C, the vitamins are destroyed.

Eating a varied and seasonal diet will help to ensure that your body gets a healthy amount of essential B vitamins. Increase your consumption of B vitamins by sprinkling salads with nuts and seeds or adding them to smoothies. Or eat an extra serving of leafy green vegetables each day.

ANTIOXIDANTS

Humans stay alive through a process called respiration, meaning that we breathe in oxygen and exhale carbon dioxide. If we could see oxygen, we would see two Os (oxygen molecules) stuck together. When we breathe, oxygen splits apart, forming two single oxygen molecules. Known as free radicals, these have the potential to damage our tissues.

One of the major ways the body defends itself from damage by free radicals is through the consumption of antioxidants. The antioxidant donates one of its oxygen molecules back to the free radical, and they pair up. Oxygen is then content again as it has its buddy back, and damage to our tissues is avoided.

Foods rich in antioxidants include all the coloured plant foods (fruits and vegetables), green tea and raw chocolate (cacao), just to name a few. This is one of the reasons why people like me bark on and on about the importance of amping up the plant content of your diet.

COENZYME Q10: PREPARE FOR IGNITION!

CoQ10 is a vitamin-like substance found throughout the body, although it is particularly concentrated in the heart, liver, kidney and pancreas – organs with the highest energy requirements (other than the brain). It is found in small amounts in meat, seafood and the oils from grape seeds and olives. It can also be consumed as a supplement, typically in the form of ubiquinol.

CoQ10's primary function is in mitochondria, the energy factories of our cells (see page 13). It's here that the food we eat is finally transformed into ATP – usable energy for our body. CoQ10 is needed for a crucial step in this process, where it transports energy-carrying electrons. It also functions as an antioxidant (by binding to damaging free radicals) in the body, and hence may improve blood vessel tone and even help to reduce blood pressure. It is particularly important for those who take statin medication, as statins can deplete CoQ10. More research is being conducted into the role that specific doses of CoQ10 can play in reducing the risk of heart disease and stroke in particular.

IRON

Iron plays a multi-factored role in energy. Adequate iron levels are critical to excellent thyroid function (see page 77) and hence energy, but iron alone also impacts on energy, predominantly via its role in oxygen transportation. Yet iron deficiency is the most common nutritional deficiency in the world. According to the World Health Organization (WHO), a staggering two billion people in both developing and industrialised countries are iron-deficient. Research suggests between 20 and 30 per cent of women of childbearing age in Australia and New Zealand are iron-deficient. This concerns me immensely as the consequences are significant, and tiredness is just the beginning.

Many people are confused about the difference between iron deficiency and anaemia. Immature red blood cells require iron to be converted by the body into a form they can use in order to mature. When fully mature, they carry oxygen from the lungs to every cell in the body. They have a big and important job to do, and this is part of the reason we feel energised when we have adequate iron levels.

Iron deficiency is the first step towards a decrease in the amount of oxygen-carrying, iron-rich haemoglobin within each red blood cell. As red blood cells are deprived of the quota of iron, they become contracted and smaller, known in medical terms as becoming microcytic. Anaemia develops when the immature red blood cells, deprived of their quota of iron, fail to survive their infancy. A formal diagnosis of anaemia is made when there is a consequent and significant decrease in the number of mature red blood cells.

Iron-deficiency anaemia can be caused by:

- inadequate dietary intake of iron
- poor absorption of iron
- loss of iron due to bleeding (menstruation or blood loss from the gut).

Heavy menstrual blood loss is a common cause, as are increased demands for iron during pregnancy. In pregnant women, iron stores have to serve the increased blood volume of the mother, as well as the needs of the growing baby. The condition can also be caused by blood loss from the digestive tract due to the long-term use of aspirin, or gastric ulcers, duodenal ulcers, bowel cancer or untreated coeliac disease.

Good food sources of iron include beef, lamb, eggs, mussels, sardines, lentils, dates and green leafy vegetables. Variety is key, as there is a small amount of iron in many foods. If you do not eat animal foods, do not assume you are iron-deficient. For some vegetarians, their body utilises the iron from vegetables sources efficiently. Vegetable sources of iron are better absorbed in the presence of vitamin C.

Iron is critical for optimal energy and long-lasting vitality. The main symptoms of iron deficiency include exhaustion, shortness of breath (especially on an incline), muscle aches and cramps, rapid pulse and heart palpitations, increased anxiety, brain fog, poor memory and concentration, headaches, depressed mood, hair loss and an increased frequency of infections. A simple blood test from your doctor will establish whether you are iron deficient or not. Testing is important, because some people have a tendency to store too much iron in the body and this needs to be avoided or treated.

If you are deficient, you will need an iron supplement to replenish your stores; diet alone won't do it fast enough for your energy recovery. It can take 18 months or longer to replenish stores if they are deficient, and by then you will have dragged yourself around and potentially damaged or interfered with other body systems for too long, adding to the recovery work that will be needed for you to get your spark back. In clinical practice I am careful which iron supplements I use, as they can cause constipation in some people. Iron liquids and some good-quality tablets don't disrupt bowel function, however, and they return iron stores to normal very effectively.

IRON DEFICIENCY IN CHILDREN

In New Zealand and Australia, it is estimated that up to 25 per cent of children under the age of three have some degree of iron deficiency – a deeply concerning statistic given that iron deficiency anaemia in early life is related to altered behavioural and neural (brain) development. Studies in infants suggest this is an irreversible effect that may be related to changes in the chemistry of the production of certain neurotransmitters (such as dopamine), and/or the organisation of part of the brain's network of wiring, or myelination (the process of forming a myelin sheath around a nerve to allow nerve impulses to move quickly). The acquisition of iron by the brain is an age-related and brain-region-dependent process, with tightly controlled rates of movement of iron across the blood–brain barrier. Dopamine receptors and transporters are altered, as are behaviours related to this neurotransmitter. It is therefore critical to a child's development that their iron levels remain adequate across all ages, a situation that is unfortunately becoming less common.

DIGESTION: THE FOUNDATION OF ENERGY

It never ceases to amaze me how magnificently clever our bodies really are – how many processes go on inside us without us having to give them any thought. Digestion is one of these, and it is central to every aspect of our wellbeing. It is the way we get all of the goodness out of our food, and the nourishment we get as a result of good digestion is an extraordinary gift without which we would not survive.

Digestion is intricate and complex, and it is intimately connected to how we feel and function every single day, from the fat we burn, to the texture and appearance of skin, to whether we have a bloated tummy. And if that isn't enough, around 80 per cent of the immune system lines the gut. So if gut function is impaired it can have a significant impact on the immune system, with more and more research suggesting that this is where some autoimmune conditions originate.

Whether your focus is optimising your health and wellbeing, improving a challenging gut, supporting your immune system, body fat management, resolving congested skin or restoring energy, understanding your digestive system is a crucial step in your big-picture knowledge of how the body works and how energy is extracted (or not) from the food you consume, allowing you to have an energised life.

The digestive system includes the mouth, oesophagus, stomach, small intestine, large intestine and numerous ancillary organs, including the liver, gallbladder and pancreas. After you chew your food it moves down the oesophagus, then through a valve and into your stomach where it is further broken down by gastric juices. The food then moves through a valve on the bottom side of the stomach and into the small intestine, through the small intestine and ileocaecal valve into the large intestine, and from there any waste is excreted. When this process works well, you look and feel fantastic. When it is in any way impaired, the opposite can be true, and correcting it can change your life.

SUPPORTING DIGESTION

Nutrients are essential for life, and the way you are eating – let alone the foods you might be choosing – may be robbing you of some of the goodness food provides, which can have an impact on every body system, including energy. Give yourself the best opportunity to absorb as much goodness out of your food as possible by applying the following tips. They may add energy to your years, and years to your life.

Chew your food

Food enters the mouth and moves down the oesophagus into the stomach. But what do we do to our food before it reaches the stomach? We chew it – or, in some cases, 'inhale' it! Many of us are in such a hurry with our meals, or we are so excited by the flavour of our food, that we might chew each mouthful four times if we are lucky. So we swallow some partially chewed food and some not-at-all-chewed food – and we do this day in, day out, year after year. And somehow we expect our stomach to cope. This alone can be the basis of problems further along the digestive tract, such as bloating.

If you are a food inhaler, try this: put food into your mouth, chew it really well, and then swallow it before you put the next mouthful in. I know that sounds simple, but try it. It can take an enormous amount of concentration for food inhalers to change their eating behaviours. Put your fork or spoon down between each mouthful if that helps. Savour the flavours, and the flavour combinations. Engage in conversation between mouthfuls if you are eating with others. Or think of your own technique to slow yourself down.

Watch portion size

Now, back to the stomach. Make a fist and observe its size. That is how big your stomach is without any food in it. Tiny, isn't it? So, think about what happens when you pile your plate high in the evening and swallow that big mountain of food. Your stomach has to stretch to accommodate it. And food sits in the stomach for a minimum of 30 minutes to allow the stomach acid and other digestive juices to keep breaking down the food.

The nerves in the stomach fire when they reach a certain stretching point, and then send a message to the brain to say you've eaten. This is one of the numerous mechanisms that has the potential to tell us to stop eating, that we have had enough. The trouble is that for some of us the stomach is so used to being stretched that by the time the nerves fire we may have already over-eaten, started to feel lousy and begun to berate ourselves.

Also, once your stomach gets used to being stretched, it expects to do so every day. This is the reason why if you decide to eat less or go on a 'diet' you tend to feel hungry after your meals for the first four days or so – it can take a while for the nerve endings around your stomach pouch to shrink back.

It is worth noting that this is how carbohydrates let us know we have eaten; with fat and protein, however, as soon as we start to chew, messages are already being sent from the mouth to the satiety centre of the brain to let us know we are eating. These signals usually reach the brain within five minutes of chewing, while the stomach-stretch method can take more like 20 minutes. This explains why it is important to include fats and/or protein with each meal, as you are likely to feel satisfied with a smaller serving.

A rough guide to the amount of food you need to eat at each meal is approximately two fist-sized servings of proteins, fats and/or carbohydrates (including starchy vegetables like potato, sweet potato and pumpkin). On top of that you can add as many water-based (non-starchy) vegetables as you like, such as spinach and other leafy greens, broccoli, cabbage and cauliflower. Essentially I want you back in charge with how much to eat for your individual needs, not relying on an external source to instruct you. So please see this fist example as a guide to portion size; the higher your muscle mass and the more you train, the more you will need to eat.

Wake up your stomach acid pH

Both the aroma of food and the action of chewing stimulate the production of stomach acid to break down our food. In the past, we took much longer to prepare our meals, and the slow cooking processes generated an aroma of the upcoming meal, signalling to the stomach that food was on its way. Nowadays, with the increased availability of fast food, and our tendency to 'inhale' our meals in front of a screen, many of us are not building up enough stomach acid.

Lemon juice and apple cider vinegar stimulate the production of stomach acid. If you haven't consumed either of these before, it is best to initially dilute them and ideally consume them five to 10 minutes before breakfast (or all of your main meals if that appeals). For example, you might begin with half a teaspoon of apple cider vinegar in as much water as you like. Over the coming days and weeks, you could gradually work up to having one tablespoon of apple cider vinegar while you gradually decrease (or not) the amount of water.

If you prefer lemon juice, start with the juice of half a lemon diluted to your taste with warm water, and gradually work up to having the juice of a whole lemon in less warm water. Or enjoy a lemon juice in a mug of warm water as your first drink for the day. If you use lemon juice, it can be a good idea to brush your teeth about 20 minutes after your meals to prevent any potential problems with tooth enamel.

Avoid drinking water with meals

We need the pH of our stomach acid to sit at around 1.9. Water has a pH of 7 (neutral pH) or above, depending on the mineral content (the higher the mineral content, the higher – or more alkaline – the pH). When you add a liquid with a pH of 7 or more to one with a pH of 1.9, you dilute it. Because we need all of the digestive fire we can muster to get the maximum nourishment from our food, avoid drinking water for 30 minutes either side of eating. If that feels impossible, leave five minutes either side of eating. Please don't wash food down with a swig of water.

Promote nutrient absorption

As food moves through the small intestine, digestive enzymes are secreted from the pancreas and the lining of the small intestine. These enzymes continue breaking down the food into its smallest components – the micronutrients and macronutrients that are absorbed into the bloodstream and transported around the body to perform their life-sustaining jobs. This is how we are nourished, and it is how we stay alive. And if this process is disrupted, inefficient or compromised in any way, you will feel like your energy underperforms every day.

However, just because you eat something doesn't mean you get all of the goodness out of it. For example, if a food contains 5 milligrams of zinc, you don't necessarily absorb the whole amount. The absorption of nutrients is dependent on a whole host of factors, some of which have been discussed already. If you 'inhale' your food, drink water with your meals, or have poor stomach acid production, for example, you may absorb very little of the goodness in your food. Fibre interferes with the absorption of some nutrients, such as zinc. Some nutrients also compete with each other, such as iron and calcium. In this case, calcium always wins, because it is a bigger structure. Some people are iron-deficient (see page 22) because they consume calcium-rich foods with iron-rich meals.

Promote good gut bacteria

Once food reaches the large intestine (colon), it is fermented by the trillions of bacteria that live there. There are countless different species of these tiny microbes living in a complex ecosystem in our gut (the microbiome), and they play an essential role not only in the health of our digestive system, but also our immune system. To give you an idea of size, you could fit a couple of million bacteria on the head of a pin!

Research in this area is (finally, thankfully!) growing, and scientists are beginning to understand how diet, medications like antibiotics and an obsession with cleanliness have potentially caused problems in the microbiome that could be linked with all sorts of diseases from obesity and type 2 diabetes to asthma and eczema. Having fewer gut bug species is a common feature of obesity, inflammatory bowel disease and other conditions. Plus, there's evidence that a diverse microbiome can better resist invasive species like *Salmonella* or *Clostridium difficile*. And what's the best way to encourage diversity? You guessed it – by eating real wholefoods, particularly as many different kinds of veggies and fruit as possible, along with legumes and fermented foods like sauerkraut.

Ensure complete bowel evacuation

When your bowel functions properly, an enormous load is taken off not only your digestive system but also your liver. The liver is one of the organs primarily responsible for minimising the build-up of harmful substances in your tissues, which is essential to optimal energy.

One of the main functions of the large intestine is to absorb water from digested food. When water consumption is low, stools become dry, hard and difficult to pass, and the longer this waste remains inside the body, the more waste will be reabsorbed back into the bloodstream. Drinking plenty of water helps promote healthy elimination and reduces the likelihood of constipation.

Incomplete evacuation or constipation could be due to a number of factors. It may be inadequate digestive processes or the insufficient production of digestive enzymes. It may be a food allergy or intolerance or undiagnosed coeliac disease. It could be related to fibre or dehydration. It can be stress hormones causing the muscles surrounding the bowel to contract and hold onto waste. It may be a magnesium deficiency that prevents the walls of the bowel relaxing and allowing the thorough passage of waste. Or the thyroid gland may not be working optimally. Traditional Chinese Medicine (TCM) teaches us that it is insufficient spleen and/or liver energy. The list of scenarios is almost endless, but finding those specific to you can change your health, your waistline, your energy and the way you experience life.

Consider food combining

When investigating gut issues in clinical practice, I start by conducting a dietary trial based on a person's history and current symptoms. This involves eliminating certain foods – the common ones are gluten, casein (dairy) and the FODMAPs foods (see page 15) – to see if the symptoms reduce. If the person responds well, that food stays out of the diet for a minimum of three

'Be responsible for the energy you bring.'

months before we try it again. If the person notices no improvement, then 'food combining' can be an exceptionally useful strategy.

Food combining produces such dramatic results for some people that I just have to share its principles. It can be a wonderful approach to eating that enhances digestion, energy and vitality, and a great way to combat a bloated tummy. Here are its main principles:

- Omit all refined sugar and processed foods, as the main goal of food combining is to initially support the health of your blood, given that it does all of the nutrient-dissemination work.
- Eat animal protein separately from starchy carbohydrates. This means you eat meat, chicken or fish with high-water vegetables (greens, broccoli, spinach, kale etc) only, not starchy vegetables such as potato, sweet potato, pumpkin and corn, or any other starchy foods, such as pasta, bread or rice (so you don't eat meat with potatoes, chicken with rice, or spaghetti bolognese).
- Eat plant-based protein (lentils, chickpeas or beans) with *any* other vegetables (water-based or starchy) but not animal protein.
- Oils and other foods rich in fats, including avocado, can be eaten with either animal-based meals or starch-based meals.
- Eat fruit at breakfast only, before any other food.

I know people who live by the concept of food combining and feel spectacular. They are brimming with energy and vitality. If this seems a bit too extreme, but you'd love to give it a go, why not use the zigzag principle? This means that most of the time you follow food-combining principles (zig), but then one day a week, or two to three meals a week, you relax the principles (zag) and enjoy whatever combination arrives on your plate (very useful if you're going out to dinner). If it makes no difference for you though, there is no need to eat this way.

Signs that your digestive system needs support:

- you suffer from reflux and/or indigestion
- you have recurring bad breath or a bad taste in your mouth
- you suffer from recurring diarrhoea, constipation or faecal impaction
- you experience intermittent bouts of diarrhoea and constipation
- you have unpredictable bowel habits
- you experience incomplete evacuation
- you have abdominal distension
- you have excessive wind
- there is an offensive odour to your flatulence
- you have pale, black or frothy stools
- you have poorly formed stools (put simply, ideally stools look like a sausage)
- there is food visible in your stools
- you have gut pain (all gut pain needs to be investigated by your doctor first)
- you have travelled and had diarrhoea (which has ceased) but you haven't felt the same since
- your tummy makes noises you can hear, particularly at night
- you burp or belch a lot; even water makes you burp sometimes
- you feel like you are becoming sensitive to foods; or are sensitive to more foods
- you rarely feel hungry (but not due to overeating)
- you feel hungry, so you start eating, but then you feel full very quickly
- you feel lousy/worse after eating
- you regularly overeat or feel 'addicted' to some foods
- you regularly take medication
- you have taken antibiotics many times
- you tend to a low mood
- you have been diagnosed with depression (80 percent of serotonin is made in the gut)
- you feel stressed regularly
- your skin breaks out easily and unpredictably, or tends to redness/inflammation
- you experience unexplained fatigue (and other causes have been ruled out).

It is important to note that this list is not exhaustive and not all of the signs listed here need to be present to be playing a role in poor energy. Please note, too, that some of these signs and symptoms can be representative of other health conditions or diseases, and that these checklists infer that more sinister causes have been ruled out by a medical professional.

Habits to better support your digestive system

- Slow down! Chew your food.
- Eat real food. Avoid processed food. Provide your body with what it has the equipment to digest.
- Include fats and/or proteins with each meal, as your energy is less likely to crash than if you simply eat carbohydrates on their own.
- If you worry that you don't eat enough vegetables for optimal health, focus on increasing your intake of plants. Wholefoods are best. You can also juice pure vegetables or make smoothies so you get the whole plant. You might prefer to use one small piece of fruit in these drinks initially for taste but aim to reduce the fruit content over time. If you travel or are busy, and making smoothies is not going to work for you, use an organic green drink powder (usually made from ground-up vegetables). The one I drink is Bio Blends Organic Daily Greens and Radiant Reds.
- Wake your stomach acid up before eating by drinking a little lemon juice in warm water or apple cider vinegar before meals, breakfast in particular.
- Drink water between meals, not with meals.
- Ensure efficient bowel evacuation. If you're not emptying your bowels every day, solve this as a priority.
- Omit a food you feel you cannot live without for a trial period of four weeks. The first four to seven days will be the most difficult, but persevere – the results will be worth it. If you omit the food (because you feel better) for longer than four weeks, you must consult a health professional to ensure you are not missing out on vital nutrients.
- Use a herbal anti-parasitic if your digestion challenges began after a gut infection. This is best guided by an experienced health professional.

STRESS AND DIGESTIVE ISSUES

Poor digestion can be due to stress or, more precisely, adrenalin, a stress hormone. When we feel panicky or worried, our body goes into fight, flight or freeze mode, and produces adrenalin, which diverts the blood supply away from digestion and concentrates it in our legs and arms so that we can escape from the 'danger'. We evolved this way for survival – if blood were still concentrated in the digestive system we might be distracted by food or hunger. Stress has a powerful impact on all our body processes, all of which impact our energy levels. I talk about ways to deal with stress on page 51. (Stress also has a significant impact on sleep, which I explore in detail on page 37). For some people, their gut does not come right until the stress is truly addressed.

WATER: THE KEY TO HAPPY, HEALTHY CELLS

Water is essential to life. Humans can survive for 20–60 days without food, depending on energy expenditure, body composition and climatic conditions, but without water, we perish within three to five days. This is because the body is around 60 per cent water and needs it for virtually every function, including circulation (blood is 82 per cent water), respiration (lungs are 90 per cent water), digestion (water helps dissolve nutrients, flush kidneys and prevent constipation), temperature regulation and joint lubrication.

Our health, therefore, depends on the quantity and quality of the water we drink. Many people do not realise that not drinking enough water can contribute to chronic tiredness, pain and inflammation. Next time you are feeling irritable, unable to focus and fatigued, think of water first. You might be surprised how far that goes to resolving – or at least improving – these challenges.

People often say they find it hard to remember to drink enough water. I suggest that they create little habits, such as starting the day with a glass of warm water with lemon juice (see page 25), or carrying a bottle of water in their bag. Drinking more water is the easiest way to improve your energy levels in the short term and to ensure your long-term health.

HOW MUCH WATER DO WE NEED?

Science currently tells us that we need 33 millilitres of water for each kilogram of our body weight. A 70-kilogram person, therefore, requires 2310 millilitres a day. Don't forget that many plant foods have a high water content (fruits and vegetables are almost always more than 70 per cent water), and this contributes to our overall daily consumption. Herbal teas and soups also add up. Foods and drinks containing caffeine and alcohol, however, draw water out of our bodies, so the larger their presence in our diet, the greater our fluid requirements.

When we exercise, perspiration and increased breathing rates mean we need more water, but the specific amounts necessary are difficult to determine and will be highly individual. Trust your thirst when it comes to this. Thirst is Nature's way of letting you know you need to drink!

Some people seem to be thirsty all the time so increasing their fluid intake is not much of a problem. For others, drinking more water makes them feel swollen and uncomfortable, or frustrated at having to go to the loo more frequently. Given these different scenarios, it's no surprise that there is conflicting advice about how to increase our water intake.

THE SCIENCE OF HYDRATION

The fact is our bodies vary in their ability to properly utilise water. In other words, you can drink plenty of water, but it doesn't mean that your cells are getting enough of it. Dehydrated cells can occur due to inadequate water intake, a lack of dietary minerals, or compromised adrenal function, often due to chronic stress (physical or emotional), excessive caffeine or alcohol intake.

To absorb water, your cells need calcium, magnesium, sodium, potassium and chloride. Some of these minerals make their home inside the cell, while others reside outside the cell wall. If the balance between these minerals is not right (they actually communicate with each other to keep the levels correct), it can be difficult for water to enter the cell. Physically, when water stays outside the cell, it manifests as a feeling of fluid retention, which, for some people, is so noticeable that clothing becomes tight as the day progresses. Improving the mineral balance of your diet and taking care of your liver is the best way to promote hydration.

If you have low blood pressure, increasing your intake of all of the minerals above can make a significant difference to dizziness, light-headedness and other symptoms. Adding a good-quality salt to food can also assist, particularly if not much, or any, processed food is consumed.

AMPING UP YOUR MINERAL INTAKE

One of the best ways to improve the intake and balance of minerals is to base your diet on what I have come to call low 'human intervention' (HI) food. Most plant foods get their minerals from the soil in which they are grown, so foods that come from organic, biodynamic or permacultured soils tend to be superior in their mineral profile. Green leafy veggies have a broad mineral profile that includes calcium, magnesium and potassium. Nuts and seeds also pack a powerful mineral punch and make a great snack or addition to any meal. Add leafy greens to as many meals as you can, and do your best to base your evening meal on vegetables, rather than them being a token effort on the side of the plate.

You can also include Celtic sea salt or Himalayan pink salt. They typically contain 84 minerals (albeit in tiny amounts) that can help your body better absorb water into its cells. Adding good-quality salt to your food can be of particular importance if you eat limited or no processed foods, especially if you suffer with digestive system problems. If you are eating plenty of the foods described above, but fluid retention is still an ongoing challenge, try juicing celery, cucumber, mint and a small amount of pineapple daily for a week. Supplementing with a 'green drink' powder made from ground-up green vegetables can also be highly beneficial (see page 29).

‘Drinking more water is the easiest way to improve your energy levels in the short term.’

ELECTROLYTES SPARK UP YOUR LIFE

There's another crucial reason to ensure a good mineral intake – your body uses minerals, among other things, to create electrolytes. Often described as the sparks of life, electrolytes carry electrical currents through the body, sending instructions to cells in all of the body systems. Electrolytes are also necessary for enzyme production. Enzymes are responsible for the biochemical processes that drive the function of the body, as well as for digesting food and absorbing nutrients, and they impact on both muscle function and hormone production as well. Poor mineral intake and/or balance, as well as dehydration, can therefore affect all body systems and functions.

SUPPORTING YOUR KIDNEYS

The primary job of the kidneys is to remove the waste products of protein metabolism from the blood. These include nitrogen, uric acid (urea) and ammonia. The kidneys also remove many other substances from the blood that could become problematic if left to accumulate, including excess hormones, food additives, vitamins, minerals and drugs.

Kidneys also help regulate healthy blood pressure and the electrolyte balance of the body, which involves the minerals needed for healthy nerve function. These include calcium, magnesium, phosphate, sodium, potassium and chloride. To give you some idea how hard our kidneys work, every 24 hours they filter about 200 litres of blood and eject 1–2 litres of waste (which leaves the body as urine). That's filtering every drop of blood in our bodies *12 times* an hour!

It is obvious, then, that if we are to have outstanding energy, or even just better energy than we have at the moment, we must take good care of our kidneys. Here are some strategies to help support their optimal function.

- Drink adequate amounts of pure water each day. Keep a large glass at your desk to ensure you stay hydrated over the day. Be happy about getting up to refill it, as this not only means you have finished a glass, you are also taking a break from sitting, which is essential to good energy (see page 41).
- Let your body guide you on how much water it needs by noticing your response to the water you drink. As you drink, observe if your thirst becomes awakened and your body actually seems to draw in the water. This is your body letting you know that it needs more water. Sometimes I start drinking and want another glass immediately. At other times, I am satisfied with one glass or a few sips. Give your body all the water it needs throughout the day to ensure that the elimination of waste via urine is well supported.
- Soft drinks, caffeine and alcohol are best minimised for good hydration. If you do consume them, drink a glass of water for each drink, as well.
- Really work to solve sleep issues, not only for the direct impact this has on energy, but also because rest and sleep strengthen the kidneys. Seven to nine hours of restorative sleep each night allows the kidneys to adequately cleanse the blood, eliminating waste products in the morning urine that would otherwise be shunted to the skin for excretion. (See the next chapter for more on sleep.)
- Take part in regular restorative movement such as t'ai chi, qi gong and yoga. These practices not only have specific poses or movements to support healthy kidney function, but the diaphragmatic breathing they foster is a powerful tool to decrease stress hormone production (see page 52).

CHAPTER TWO

ARE YOU SLEEPING WELL?

Sleep, and the rest and repair it offers, is critical to life. Restorative sleep ensures the optimal functioning of our bodies – our physical endurance, reaction time, mood, memory, cognition, digestion, sex-hormone balance, thyroid function and even the condition of our skin. If you have poor sleep quality or do not get enough sleep, you can wake up just as exhausted as when you went to bed – something that's very common, but not normal, as I like to say.

Many people blame age when they start to feel more and more tired as the years go by, but it doesn't have to be this way. If low energy truly was down to age, then every 82-year-old would be exhausted, and they're not! You can make a really big difference to how you feel and function, both inside and out, by ensuring you get enough good-quality sleep.

So how much is 'enough'? It depends on your age, gender and aspects of your biochemistry, but adults generally need between seven and nine hours sleep each night in order for the body to complete all of its vital cellular repair work. When cells are working optimally, we look and feel our best.

Yet studies show that people today are simply not getting enough sleep, let alone good-quality sleep. For example, in New Zealand in 1960, people reported that they were getting about eight-and-a-half hours of sleep each night. By 2002, the number had dropped to six-and-a-half hours, and it is still falling. Other Western countries report similar findings. This reduction in sleep affects our health, our longevity, how we feel, how we function, the patience and kindness we display, and how energised we are each day. And given that how you feel doesn't end with you – it has an impact on every single person you come into contact with – the ripple effect of lousy sleep can create additional challenges in your relationships at home, at work and with friends.

While many people compromise their sleep to get more done in a day, others make an enormous effort to get plenty of sleep but their bodies won't cooperate.

OUR BODY CLOCK

Circadian rhythms are important in determining human sleep patterns. The 'master clock' that controls them consists of a group of 20,000 nerve cells in the brain called the suprachiasmatic nucleus (SCN). The SCN is located in the hypothalamus, an area of the brain just above where the optic nerves from the eyes cross. The SCN controls the production of melatonin, one of our sleep hormones. Since it is located just above the optic nerves, which relay information from the eyes to the brain, the SCN receives information about incoming light. When there is less light, which is what is supposed to occur at night, the SCN tells the brain to make more melatonin so you get drowsy. Trouble is, too many people spend each night exposed to too much light, too late into the evening. And when we live our lives in this way, where our surroundings continually communicate messages to our internal assessors (essentially, our endocrine system), the cascading effect of those signals can disrupt our most basic and essential of rhythms.

TIPS FOR RESTORATIVE SLEEP

When my clients tell me they don't sleep well, this usually means they have trouble falling asleep, or they wake during the night and have trouble going back to sleep. Or they sleep all night, yet wake up feeling just as tired as when they went to bed. There are many factors that interfere with restorative sleep, including stress hormones and living with the fight or flight response switched on, digestive issues and too much light too late at night (think screens), but I want to share some simple sleep ideas to help you feel more rested.

'Great-quality sleep is critical to outstanding energy.'

GO TO BED EARLIER

This one sounds simple, but many people will claim that they are unable to get to bed earlier. If you are someone who stays up later and later in an attempt to get more things crossed off your 'to-do' list, it may take an extreme question for you to follow through with making sleep a priority. Try asking yourself: 'Will anyone die if I don't get this done?' If the answer is yes, then stay up later and do it. If the answer is no, *go to bed*! Some of you will have to go to this extreme to make sleep a priority. But I say this to you to bring home just how critical great-quality sleep is to outstanding energy. You will do whatever you prioritise.

USE SOFT LIGHTING

Melatonin is your primary sleep hormone: it helps you fall asleep and stay asleep, and its production in your body is disrupted by light. So if you expose your eyes to light too late into the evening through any means, including the use of back-lit devices such as mobile phones and laptops, or you work until late under bright light or watch TV, it can stop your body from producing the very hormone that is necessary for great sleep – and hence great energy.

If you don't wake up refreshed, or particularly if you have trouble falling asleep, become very aware of how much light your eyes are exposed to within two hours of bedtime.

Do what you can to only be in soft light in the lead-up to bedtime, and make your bedroom a screen-free zone. Street lights can also disrupt sleep. Black-out curtains might be the best investment you make for your sleep, while others prefer to let the sun wake them in the morning.

ENERGY-ZAPPING SCREEN TIME

As you now understand from the earlier discussion about circadian rhythms, light can disrupt the messages the body is supposed to receive to wind down and fall asleep.

Concerns have long been raised about the effects of excessive TV viewing, including reduced energy levels, poor concentration and the health implications of lack of exercise. However, with the rise of digital media, more and more people are not only spending their working days in front of computer screens, but also their leisure time on smartphones, tablets or laptops. An Australian study in 2014 found that 80 per cent of the 1500 adults surveyed spent an average of 25 hours a week online – the equivalent of a whole day a week – and that didn't include any time in front of the TV.

Most people are oblivious to the amount of time they stare at a screen, so it can be really helpful to keep a screen diary for a week. Make a list of what you watched and for how long. Also consider your reasons for watching TV or surfing the net. Is it because you have had a stressful day at work and you just want to switch off your brain? Or is it because you feel bored or lonely? If the latter, why not go for a walk, phone a friend or relative, cook something delicious for family snacks or lunches, or plan a weekend away. By engaging in more active or restorative tasks, you may notice your energy levels increase, along with your feelings of joy and contentment.

ADJUST YOUR SLEEPING ENVIRONMENT

A bedroom that is overly warm or cold can disrupt sleep. Notice the temperature that best serves you. Make sure your bed is comfortable, too. People often keep their beds for far too long, so are not getting enough structural support to encourage a restful sleep. Sometimes you might wake in the night from pressure points in your shoulders or hips making you uncomfortable. If the last time you slept well was in a hotel (the beds tend to be newer), it might be time to upgrade your mattress.

AVOID HEAVY, SPICY FOODS

You might like to try avoiding heavy or spicy food in the evening or eating smaller portions, as this may positively impact on your sleep and help you wake more refreshed. Eating earlier can also help you have better-quality sleep.

AVOID ALCOHOL

This sounds counterintuitive, as alcohol is a depressant and seems to help us fall asleep quite quickly. However, the reality is that alcohol disrupts rapid eye movement (REM) sleep, the fourth part of the sleep cycle, which is when the body undertakes critical repair work. Alcohol can also lead people to wake up hot, typically between the hours of 2 am and 4 am. Notice how alcohol affects your sleep, and consider making changes. Alcohol also has to be converted by the liver into another substance that can be excreted (the body can't eliminate alcohol), and when the liver is in overdrive from the regular over-consumption of alcohol, it can also wake us from heat.

GET UP AT THE SAME TIME EACH MORNING

Another powerful way to help reset your body's own natural circadian rhythm is to get up at the same time each morning and expose your eyes to light. Preferably get up and go outside and exercise. If that is not practical for you – perhaps because you have young children who need you – then, on waking, get up and fling the curtains open and notice the day and Nature outside. Use this ritual to remind yourself to think of three things you are grateful for, allowing your eyes to gently be exposed to the light of this new day while you do. Commit to doing this for a week, and notice if this ritual begins to make a difference to your quality of sleep and your energy.

OTHER FACTORS THAT INTERFERE WITH SLEEP

The following factors may also interfere with your ability to sleep. Reflect honestly on your life, and see if there is something here that needs exploring or support.

WORRY

From time to time, worrying can play a role in being able to sleep well. You might worry about paying the bills or about challenges your child is experiencing. A loved one may be ill. You may be someone who worries about what others think. You may worry about things you have said or done over the day and what others said to you. You may be striving to please everyone in your life, and so you worry you may have let someone down.

What can you do about these things when you are lying in bed in the middle of the night? Nothing. In that restless moment, accepting that there is absolutely nothing you can do to rectify a situation that may not even happen can be a great start. My dear mum gave me a wonderful piece of advice about worrying. She suggested: 'Don't worry about anything until it is a problem.' That is such good advice. You can take it one step further if you like. Even if something is a problem, worry still won't solve it. Breathe. And simply decide to face the situation or the person with your authentic self and be present in that moment. Life asks this of us and we get to choose if we front up or turn away. Showing up leads to resolution, less worry and better sleep.

If you are churning out stress hormones and they are communicating to every cell in your body that you are not safe, your body does not want you to sleep deeply, as you need to be able to wake up quickly and save your own life. Your body has your best interests at heart. You just need to communicate to it that you are safe and that it is safe to sleep deeply and restoratively.

As I explain on page 52, there is no more powerful way to calm your nervous system than through diaphragmatic breathing. I know it might sound too simple to make a difference, but how you breathe consistently over a day – short, sharp, shallow breaths driven by adrenalin, or long, slow, tummy-moving diaphragmatic breaths – has an impact on your nervous system, blood chemistry, oxygenation and hence your energy.

You can also help your body produce fewer stress hormones by decreasing, or omitting, caffeine, and beginning to identify where you perceive pressure in your life where you don't need to. These shifts can help you resolve your sleep challenges so that you wake with better energy. After all, historically and still today, stress hormones communicate to the body that your life is in danger; therefore, sleeping deeply risks your survival – hence why stress hormones are so effective at interfering with good-quality, deep, restorative sleep.

MAGNESIUM DEFICIENCY

Magnesium and other minerals are critical for the body to physically relax, particularly the muscles. However, many people today don't consume enough. The body's magnesium requirements fluctuate, too. When we are stressed and producing adrenalin, each unit of adrenalin that is made utilises extra magnesium. This can lead to the muscles not having enough for sufficient relaxation. For good-quality sleep, include optimal amounts of magnesium in your diet, found predominantly in green leafy vegetables, nuts and seeds. Some people's sleep and energy improve by taking a good-quality magnesium supplement.

SLEEP APNOEA

The widely accepted definition of obstructive sleep apnoea (OSA) is a clinical condition in which there is intermittent and repeated upper airway collapse during sleep. This results in irregular breathing at night and excessive sleepiness during the day. It can be a major cause of fatigue.

In many people with OSA, airways become blocked due to excess body weight and a build-up of soft fatty tissue in the windpipe. It is estimated that about 70 per cent of all people with sleep apnoea are overweight. However, other issues can also contribute to sleep apnoea, such as enlarged tonsils, a deviated septum or small or narrow airways in the nose, throat or mouth. This narrowing can be structural or due to allergies that have never been diagnosed or treated, with the most common culprits being dust mites, animal hair and casein in dairy foods. It can also be due to an inability to retain adequate carbon dioxide in the blood, as carbon dioxide is a vasodilator, meaning one of its actions is to broaden blood vessels, thereby allowing more oxygen to be delivered throughout the body.

People who smoke and have high blood pressure are also at a higher risk of OSA, and it is more common in men. Interestingly, neck circumference can be a strong predictor too. People with a neck circumference of less than 37 centimetres are considered to be at low risk, whereas those with a neck circumference greater than 48 centimetres have been found to be high-risk. An ear, nose and throat specialist, a GP or a dentist may assess for the presence of nasal polyps, rhinitis or any deformity of the nose to determine the cause of any obstructions. Getting to the heart of sleep apnoea or low blood oxygenation and resolving any issues can make a world of difference to fatigue.

Your medical professional can conduct sleep studies, and if the descriptions opposite resonate for you or a loved one, this is well worth investigating. The treatments are varied, but all aim to ensure that adequate oxygen is entering the body and being delivered everywhere it needs to go, a process essential for optimal energy.

SYMPTOMS OF SLEEP APNOEA

Since many sleep apnoea symptoms occur during sleep, it can initially be challenging to detect the disorder. One of the most common signs is loud, chronic snoring, often followed by choking or gasping. As sleep apnoea progresses, snoring may increase in volume and occur more frequently. However, not everyone who snores has sleep apnoea. Other symptoms include:

- daytime sleepiness
- morning headaches
- difficulty concentrating
- memory problems
- regular, repeated night-time urination
- irritability
- mood swings or symptoms of a depressed mood
- a dry throat on waking.

CHAPTER THREE

ARE YOU MOVING ENOUGH?

The human body is designed to move, and exercise is important for a whole range of reasons, including strength, flexibility, body shape and size, bone density and mood, just to name a few. Building muscle is also critically important to – among other things – increasing and enhancing mitochondrial function. As we saw in chapter one, mitochondria are the power plants in the cells (the 'mouse on the wheel') and are in almost every cell type and tissue in the body, from the brain to the thyroid gland to the tendons around the knees. In short, trillions of mitochondria are distributed throughout the body with the sole purpose of generating ATP (the cellular form of energy).

In order to provide the large amounts of ATP needed for movement, muscles contain the highest mitochondrial content of any tissue in the body. Each muscle cell can contain hundreds or even thousands of mitochondria so that they can support the generation of large quantities of ATP during exercise. The other amazing thing about mitochondria is that they can replicate themselves. Certain types of exercise can induce large increases in muscle mitochondrial content, and thereby increase the ability of muscles to take up glucose during and after exercise, an important process known as mitochondrial biogenesis.

Are you starting to get a sense of how important it is for you to maintain (or preferably build) muscle mass for optimal energy? Remember, muscle mass naturally declines from about the age of 30 onwards unless you do something to counteract this. Sedentary lifestyles and chronic disuse of muscle result in a decline in mitochondrial numbers and function, leading to the production of free radicals and cell death. Think about that. If a cell dies, you lose a mouse. And that means you lose energy-producing capacity.

The muscle tissue of people with type 2 diabetes has been extensively studied, revealing gross defects in mitochondrial number and function. Although the cause and effect remains unknown, muscle tissue from people with type 2 diabetes is often associated with reduced aerobic capacity, insulin resistance and deficient mitochondrial biogenesis. Studies have also shown that defective mitochondrial biogenesis in the heart can predispose individuals to cardiovascular complications, heart disease and metabolic syndrome.

Fortunately, reversing the effects of ageing, type 2 diabetes and cardiovascular disease via increased mitochondrial biogenesis is as simple as exercising more. Truly, if exercise were a pill, everyone would take it! Studies have shown that if aged individuals with existing metabolic disease resume an active lifestyle, they can significantly improve pre-existing cellular damage and promote gains in muscle mass, as well as activating mitochondrial biogenesis.

So the next time you exercise, imagine the work that your mitochondria are doing. I assure you, their behind-the-scenes effort is nothing short of magical.

HOW SITTING DRAINS ENERGY

It may surprise you to learn that even if you do exercise regularly, the amount of sitting you do during the day may well counteract the benefits to your energy. Studies show that some of us spend 11 hours or more per day on our bottom: commuting to and from work, sitting at a desk all day, then watching TV or sitting at a computer for the rest of the evening.

Numerous research groups have studied the physiological consequences of prolonged sitting. They suggest that the loss of local contractile stimulation induced through sitting leads to the suppression of skeletal muscle lipoprotein lipase (LPL) activity. LPL is necessary for triglyceride (free fat in the blood) uptake and HDL-cholesterol production, as well as reduced glucose uptake.

In other words, when we sit for too long, the body's ability to burn energy is compromised, which can promote weight gain and lethargy. Think about it. If you sit for five hours straight, do you jump up enthusiastically afterwards or do you begin to droop over the arm of the chair or your desk and feel sluggish and fatigued? Typically, it's the latter.

GET MOVING!

Fortunately, research has shown that frequent, small amounts of movement can improve some blood markers for cardiovascular disease and type 2 diabetes. Plus, it can help clear some of the inflammatory products from the body, reduce back, neck and shoulder pain and also boost mood – all factors that contribute to us experiencing better energy, both physically and emotionally.

To reap the health and energy benefits, it is best to get up (from your desk) every hour for about three to five minutes. You might like to walk up and down a flight of stairs a few times or walk the length of the office. Or get yourself a glass of water from a location distant from your desk. You could even invest in a standing desk or create one from objects around your home or office.

FIVE WAYS TO MOVE MORE

- Ride or walk to work or school, or get off the bus/train/tram a couple of stops earlier so you have to walk further.
- Swap sitting meetings for walking meetings at work: use your phone to record the meeting then sit together for five minutes at the end to consolidate.
- Take the stairs instead of the lift or escalator.
- Make family time active: play hide-and-seek, backyard cricket or walk to the park instead of watching TV.
- Get up during TV commercials and go for a walk around the house.

POSTURE AND ENERGY

Although the way you hold your body is not about movement, it is still a physical aspect of being that affects our energy levels.

Try this exercise. Slump your shoulders. Let your head hang forward and tilt your face down. Hold that posture and stand or walk around like that for a few minutes. How do you feel? Now pull your shoulders back, hold your head up, open across your chest and face forward. Spend a few minutes like that.

Which posture offers you more energy?

The upright posture, of course! So try to become more conscious of how you hold yourself. Make it part of how you care for yourself every day. Dr Donny Epstein, the genius who created Network Spinal Analysis (NSA), teaches that over time, due to the way our nervous system interprets the experiences we have – a raised voice, a failed exam, a car accident, the passing of a loved one – we develop defence postures. Each posture we have offers us a concurrent headspace, and the less defensive our posture is the more resources (from within), including energy, will be on offer to us.

Posture significantly impacts energy. Stand up tall and lose the weight of the world off your shoulders. Different emotions will likely be present in each different posture, even though you may not link them to how you are holding yourself. An NSA practitioner can take you further on this journey if that appeals or is needed.

'Truly, if exercise were a pill, everyone would take it.'

CHAPTER FOUR

ARE YOU STRESSED & EMOTIONAL?

Human beings have always experienced periods of stress; it's nothing new. But a relentless, constant, urgent avalanche of it has become the norm for too many people. We worry about relationships, finances, our health, the health of our loved ones, the size of our thighs, the future of the planet, or simply whether we've upset someone or not. We feel pressured about our emails, about work, about our bodies, about getting things done, about never letting anyone down – yet feeling that whatever we do, it is never enough; we aren't enough.

And while you believe that you aren't enough, you will never feel like you have enough, and the relentless pursuit for 'more' continues, exhausting you further. You crave a coffee in the morning to get you into your day, then wine in the evenings to help you relax, even though underneath you are utterly exhausted. Many people have become so accustomed to living this way that they don't even notice how stressed they are anymore.

HOW OUR BODIES RESPOND TO STRESS

When we are stressed, our adrenal glands secrete the hormones adrenalin and cortisol. The adrenal glands are part of the endocrine system, and sit just on top of the kidneys. They also produce sex hormones such as progesterone, as well as hormones that help control blood pressure, fluid balance and salt retention, among other things.

Adrenalin and cortisol behave very differently from one another, and communicate different messages, but both have the power to influence whether we bounce out of bed full of vitality in the mornings or can barely reach over to turn off the alarm. Stress hormones don't just impact energy systems directly. They can also interfere with other processes that are vital to metabolic rate and energy levels, such as blood glucose regulation, immune response, thyroid function, gut function and sex hormone balance. They impact whether the body is getting the message to burn body fat as a fuel or store it. Stress hormones and the body systems they impact can also influence whether we feel calm or anxious, happy or sad – emotional states that in turn influence food choices and our level of self-care and self-esteem. Every physical system in the body is interconnected and influenced, in turn, by a person's mental and emotional state. There are ripple effects in action everywhere.

THE AUTONOMIC NERVOUS SYSTEM (ANS)

The autonomic nervous system (ANS) is the part of our nervous system that is not under our conscious control. It regulates our heart rate, respiration rate, temperature and immune and hormonal systems while we carry on with our lives. It's the part of us that heals a cut. We don't have to instruct it – in fact we can't instruct it – it just happens.

There are three parts to the ANS: the sympathetic nervous system (SNS), the parasympathetic nervous system (PNS) and the enteric nervous system (ENS). Here I will focus on the SNS (the 'fight, flight or freeze' response), and the PNS (the 'rest, digest, repair, reproduce' arm of the nervous system) and their effects on health and energy.

When our brains register a threat to our safety (whether imagined or real), the SNS raises our heart rate, increases our respiratory rate, releases stress hormones, and shunts blood away from the digestive tract to the muscles so that we can deal with the threat by running, attacking or playing dead. It also downregulates our reproductive system (as there's no point procreating when our lives are threatened). If organ systems in the body are unhealthy, or if we are mentally or emotionally stressed, this also increases the sympathetic load.

Once the 'threat' is dealt with (although is it ever dealt with in the modern world?), the PNS kicks in to slow our heart rate and respiration and return blood to the digestive tract so that we can digest our food. It also allows libido to be restored.

The challenge for so many people's energy today is that they live in a state of SNS dominance. This means their body is getting the message, via their nervous system, consistently, that their life is literally in danger.

Science suggests that humans have been on the

planet for 150,000 years. In your mind, imagine that length of time represented as a 30 cm school ruler. To put the era we are now living in into perspective, 500 years is 1 mm on that ruler. So the last 20 years is a drop in the ocean of time that we have been on the planet. Think about it. *So* much has changed in the *very recent* past. Social media is just over 10 years old. It wasn't that long ago that when you left the house, no one could contact you. And the food supply is dramatically different from what we have evolved eating. Processed foods are very new to our bodies.

The way we now live is *very* new to us as a species. For example, when you look at time in this way, it wasn't that long ago women were given the opportunity to do what had traditionally been their father's work, while maintaining what were traditionally their mother's responsibilities. So what has unfolded for too many women is a frantic double shift of work day and night, with very little, if any, rest. And we can't fight our biological need for rest. Women and men both need to adapt to new ways of living.

When we live constantly in SNS dominance, it is essentially adrenalin and cortisol – our two stress hormones – driving this. The female body links another hormone called progesterone to fertility. Progesterone also plays a number of other roles in the body such as acting as a powerful anti-anxiety agent, an anti-depressant and a diuretic (helping to get rid of excess fluid). Yet when adrenalin is communicating to your body that your life is in danger – as that is what it has meant for 150,000 years – and cortisol says there is no food left in the world (see page 49) the body thinks it is doing you a great big favour shutting off your adrenal supply of progesterone, for it doesn't want you to bring a baby into a world where it perceives you aren't safe and there is nothing to eat . . . This can lead to challenges with fertility, anxious feelings, low mood and/or fluid retention. I did a TEDx talk about this, as I believe it is the biggest health challenge facing women in the Western world today.

It is an essential life skill today for us to have daily practices that activate the PNS. This means doing what we can towards less SNS activation – minimising caffeine and exploring our perception of pressure and urgency and saving it for when we really need it.

Don't make what you have to do each day full of stress and pressure. Don't lose touch with how privileged your life is because all of your basic needs are met, when for too many people in the world this is still not the case. See your days as busy, full of opportunities, rich with experiences. I'm not denying there is not real stress and trauma; sadly, there is. But your email inbox and your to-do list do not fit into that category. Your *perception* makes them that way. You also want to activate the PNS, and diaphragmatic breathing is the only way we currently know that does that, specifically extending the length of the exhalation. Remember, you can't boss your ANS around. You can't tell yourself to chill out when you have surges of adrenalin coursing through your blood because you've had a few coffees and unexpected deadlines this morning. The only way you can impact your ANS to activate the PNS is via how you breathe. Because you'd never be able to breathe diaphragmatically if your life was truly in danger.

Living in SNS dominance influences almost every body system. It impacts everything, from digestion to sex hormone balance, food choices to whether you predominantly use body fat or glucose as a fuel, sleep to thyroid function.

You don't have an illness or disease. But you're not okay either. You no longer feel like yourself and on some level you're heartbroken about this . . . not that you slow down long enough for this to sink in. Your energy is like a rollercoaster: caffeine warms you up in the morning and at the end of the day, you are tired but also wired so you feel like you need alcohol to calm you down. This can turn into a vicious cycle, never really addressing the cause of your poor energy – SNS dominance.

ADRENALIN – THE ACUTE STRESS HORMONE

Adrenalin is the short-term, acute stress hormone, designed to get us out of danger fast. It elevates heart rate, blood pressure and respiration in readiness for action, and mobilises the body's glucose stores. The liver and muscles store glucose in the form of glycogen; when adrenalin signals the need for energy, the glycogen is converted back into glucose, and released into the bloodstream ready to fuel self-defence or escape.

Adrenalin communicates to every cell of your body that your life is in danger and it prepares you to fight or flee, which was useful in the past if a tiger suddenly jumped out at you, or a rival tribesman started chasing you with a spear. However, these days, you may be making adrenalin simply because you have to make a phone call that you would rather not, or perhaps because you have already gulped down three coffees and it's not even lunchtime. Or maybe your dad yelled at you a lot when you were a child, and even though you know he loved you and that it was just his way of coping, whenever a male raises his voice now you instinctively go into fight-or-flight mode. The majority of stress for most people in the Western world today is psychological rather than physical, and it can be constant and relentless.

With adrenalin production comes glucose mobilisation – in other words, blood glucose levels increase to fuel our fight or escape. The trouble is, we're not burning up all of that glucose – we're often just sitting at our desk, or in our car, our hearts thumping, our thoughts racing and our muscles tensed. So we have all this unused glucose surging through our blood, which then triggers the release of insulin whose job it is to gather up all the excess glucose and convert it first to glycogen and, when the glycogen stores are all full, to body fat. When you do this day after day, year after year, it can be one of the roads to insulin resistance and all of its dire consequences, which includes type 2 diabetes. This can also lead to the body's inability to 'hear' the appetite-regulating signals of leptin, as high circulating insulin levels interfere with the instructions of leptin that let you know you can stop eating. This in turn prevents the body from efficiently converting thyroxine (T4), the inactive thyroid hormone, to tri-iodo-thyronine (T3), the active, metabolism-driving, energy-boosting thyroid hormone. Instead, you make reverse T3 (rT3), which doesn't have the same positive impact on energy and metabolism; it makes you feel deeply fatigued, tired in your bones.

So you can see that fatigue can be generated by a cascade of biochemical processes not working optimally. Nothing in the body stands alone, and understanding what has created your fatigue is essential to being able to guide you out of it.

STRESS & SUGAR CRAVINGS

When we live in SNS dominance, it leads us to crave sugar, and here's why. When the brain perceives a threat to survival, the body requires an injection of fuel to get you out of danger, and it just so happens that glucose is the most efficient fuel (body fat takes longer to burn). So even though you know that eating sugar doesn't serve your health, you still crave it. A 70 kilogram adult is able

'Notice if what you just said, did or ate fosters energy or depletes it, in you and for those around you.'

to store about 2500 calories of glucose (it's stored as glycogen in the liver and muscles), and about 130,000 calories of fat. However, glucose is your 'get out of danger' fuel and when you use it constantly, because of all the coffee you drink and a to-do list that is never all crossed off (short, shallow, upper chest, adrenalin-based, SNS-style breathing), your body automatically detects when it's getting low and switches on your craving for sugar. This is a survival strategy that ensures you always have enough fuel to power yourself out of danger! So you crave it. Intensely. And you usually eat it. I'm not saying willpower or a decision to take better care of yourself can't play a role and make a difference. Of course it can. We can all decide in this very moment that our body is precious and worth taking care of and that rubbish no longer belongs in there. We can all do that. I simply want you to understand that there are ancient, biochemical mechanisms going on inside you all day every day, which drive behaviours and choices in your life that you may not understand. Living in SNS-dominance is one of them – and one of the biggest reasons that so many people have lousy energy.

CORTISOL – THE CHRONIC STRESS HORMONE

Cortisol is the long-term (chronic) stress hormone that kicks in when you've been making adrenalin for too long. Physically, the adrenalin increases inflammation and one of the positive roles cortisol plays is to dampen down that inflammation. However, too much cortisol has a myriad of negative effects that impact body-fat levels, muscle mass, metabolism, blood glucose, immunity and energy.

In evolutionary terms, long-term stress for humans revolved around food scarcity – floods, famines and wars. During such times, a person didn't know where the next meal was coming from. Today our long-term stresses tend to be based more on financial situations, relationship concerns, challenges with friendships (particularly for teenage girls), and uncertainty or worries about our health or body. I can't tell you how many times I have heard women, in particular, say that they would 'do anything' for a different body part – thinner thighs, less body hair, no cellulite – and for many women the part or parts of themselves they dislike intensely become a silent fixation in their minds. For many women, their first waking thoughts involve 'What will I, or won't I, eat today?' or 'How much exercise can I get done today?'

So many people zoom through their days with a pervasive not-good-enough or never-enough-time monologue running through their head. Day after day, this can easily lead to a chronic pattern of stress response, hence increased cortisol output. In turn, this can lead to a significant change in your metabolism, your energy and where you store fat on your body.

LONG-TERM STRESS & WEIGHT GAIN

Cortisol is not a 'bad' thing. When it is produced in the right amounts, it not only acts as an anti-inflammatory but also buffers the effect of insulin, helping you burn body fat for energy while also maintaining stable (as opposed to rapidly fluctuating) blood sugar levels.

However, when cortisol is elevated over a long period, it tells every cell of your body that food is scarce, and that you must slow down your metabolic rate and store fat in order to survive the 'famine' your body is expecting.

One of the ways cortisol slows down your metabolic rate is by breaking down your muscles to use for fuel (with less muscle mass you burn fewer calories). The protein is broken down into amino acids, and the amino acids are in turn converted to glucose, which your body thinks may be useful to assist you in your stress. Yet, if you are not active, this increase in blood glucose will not be utilised, and insulin will have to be secreted to put the excess glucose back into storage. (Remember that glucose is stored as glycogen in the muscles and the liver.)

However, because your muscles have been broken down, there is actually less space for the glycogen to be stored, so it has to be converted to body fat. (This is also one of the mechanisms that may create cellulite, as fat is deposited in places where you used to have muscle.) This is the process through which long-term stress can contribute to type 2 diabetes.

Cortisol has a distinct fat deposition pattern. If your cortisol levels are chronically high, you will typically develop fat around your tummy – a pattern governed by the body's quest for survival (if food suddenly did run out, your major organs would have easy access to fuel to keep you alive, plus the warmth and protection extra body fat on the torso would offer for this 'famine'). You also tend to lay down fat on the back of the arms ('bingo wings'), and you grow what I lovingly call a 'back verandah'.

So what do most people do when they notice that their clothes are getting tighter? They go on a diet, which typically means they eat less. Eating less, though, just confirms to your body what it perceives to be true – that there is no food left in the world – and so your metabolism is slowed even further.

This can lead people to feel that their body is betraying them, for even though they may be making great efforts with food and movement, they are not achieving weight loss as a result. If anything, their clothes continue to fit them the way they have for a while or they become tighter, despite great efforts. This alone can be incredibly disheartening. Going on a diet is never the right medicine.

HOW CAFFEINE ADDS TO STRESS-HORMONE LEVELS

When clients talk about feeling stressed, overwhelmed and anxious, one of the first things I ask is whether or not they drink a lot of caffeinated drinks. Caffeine acts on the adenosine receptors in the brain, which signal to the adrenal glands to make adrenalin, driving a plethora of metabolic consequences related to the body's perception of safety, such as impacts on blood pressure, whether your tummy is bloated or not, and whether you are able to effectively use body fat as a fuel or not.

So many people run on adrenalin these days. And it doesn't have to be triggered by a traumatic event or shocking situation. It can simply be the pace at which we live our lives: being contactable 24/7; constant exposure to social media, unless we purposefully choose otherwise; the juggling act that leads so many people I meet, women in particular, to say that they want more 'balance' in their lives, as they can't cope like this anymore.

So if you are already experiencing a lot of stress, consider your caffeine habits and get honest with yourself about how it affects you. Does it dull your appetite, so that you tend to grab a coffee instead of eating a proper meal? This is especially true for many women at lunchtime. Does it make your heart race, give you the shakes, or loosen your bowels? Does it elevate your blood pressure? Do you have restless, poor-quality sleep because of how much caffeine you consume? Do you notice you want it more when you are stressed? Does it lead you to snap at people and then you wonder afterwards why you reacted so intensely to something so (relatively) small? Or does it lift your mood, help you think straight, or nourish your soul with no ill effect whatsoever?

I have asked many of my clients to decrease their caffeine intake or take a break from it completely for a four-week trial period, and they are often shocked by how much more energy they have without caffeine in their lives, not to mention having less anxiety! You know yourself better than anyone. Act on what you know is true for you.

> **A LITTLE NOTE ON TEA**
>
> Not so long ago, before the coffee culture hit town, tea was the major source of caffeine for people in Australia and New Zealand. Tea contains caffeine, although how much depends on how long you brew it. An average cup of black tea contains 50 milligrams of caffeine, whereas green tea tends to contain about 30 milligrams per cup. Tea, however, also contains another substance called theanine, which acts as an antioxidant and also helps buffer the effect of the caffeine. It makes you alert but not wired. Some people are actually highly caffeine-sensitive, and even green tea can impact on them negatively. However, many people tolerate tea and feel good drinking tea compared with coffee. They may not realise this until they take a break from coffee, though. Do what serves your health and energy.

OTHER EFFECTS OF CHRONIC STRESS

When levels of stress hormones are high, the body puts all of its resources into saving your life rather than into what are considered non-vital processes, including digestion, immune function and the nourishment of skin, hair and nails. Reproductive functions are also down-regulated, since they use a lot of energy and are not necessary for your immediate survival, given the impending 'threat'. Plus, your body does not believe it is safe to bring a baby into what it perceives to be an unsafe world, as adrenalin is telling your body that your life is in danger, while cortisol communicates that there is no more food left in the world!

Over time, the lack of resources (blood supply, oxygenation and nutrients) available to these non-vital processes (i.e. those not necessary to sustain your life) has significant consequences internally and externally, *on top of* the consequences of the stress hormones themselves.

WAYS TO ADDRESS YOUR STRESS

So how do you support your nervous system and stress-hormone production for great and consistent energy?

EMBRACE A RESTORATIVE PRACTICE

If you feel that stress is really at the heart of your low energy, then I urge you to consider this. Yes, stress can be incredibly energy-zapping, particularly chronic stress. Yet the real word for stress is fear. Whatever you are stressed about – anything at all – it is usually what you are frightened of. Peel back the layers of what you perceive are your stresses and see what's really there. See what you are actually afraid of – being seen as lazy, feeling like a failure, people not liking you, letting others down . . . For most people, when they peel it all back, their fear is that they are not loved, or that there will be a loss of love. (Swap the word 'love' for 'acceptance' if that feels more comfortable to you to get the insight here.) Everything – and I mean everything – comes back to our drive to avoid rejection and obtain or maintain love. I don't know how else to say it. People think the opposite of stressed is relaxed or calm. I say it is trust.

You have inside you 'a natural intelligence', as Marianne Williamson so beautifully shares. You can name this what you like, but for now let's call it that. An acorn grows into an oak tree. What does that? A human doesn't instruct that and make that happen. The acorn has a natural intelligence that grows it from its baby self into the highest, best version of itself as a beautiful oak tree. Same goes for a rose. Do you instruct the rose to open? No, its own intelligence grows and opens it. Do you really think you are separate from that? You have the same natural intelligence inside you, wanting to grow you to become your highest, best version of yourself. Natural intelligence wants nothing more for you than to be your best version of you. The thing is, most of us walk around as though we are lamps, busting to shine our light, but we aren't plugged in. We spend our time in houses where the electricity is connected, but our light never goes on. In your heart of hearts you know if this is happening. You sense there is more to life. You sense you have potential that you haven't yet tapped. And this gnaws away at you. You have boundless energy when you are plugged in. You trust. You trust the unfolding of life; you stop needing to control everything. You believe that life doesn't happen to you, but that life happens for you. So how do you plug in? The only way I know is meditation. That allows you to hear the natural intelligence that is always there, within you, waiting for you to listen.

Diaphragmatic breathing

If meditation isn't your cup of tea, or if yoga, t'ai chi, qi gong or any other restorative practice isn't something you're interested in, then I urge you to consider diaphragmatic breathing itself as a daily practice. It truly is a game-changer – perhaps even the single most important thing to take away from this book.

If you watch the way people breathe, you'll notice that many breathe from the top of their chest – this is the only part of them that moves in and out. Adrenalin drives those short, sharp breaths. If you breathe in a shallow way, with short, sharp inhalations and exhalations, then you are communicating to your body that your life is in danger. It then has to make adrenalin. Remember the cascade of hormonal events that follow such an alarm, and the role these hormones play, for one thing, in switching fat burning on or off, and depleting energy?

How you breathe can also be a fast-track to feeling anxious, and potentially panic attacks, regardless of what led you to breathe in a shallow way in the first place. Whether it was an event, a deadline, caffeine, the perception of pressure, the 'need' to rush, or the lifetime habit of your nervous system, the result is the same.

Long, slow breaths that move your diaphragm communicate the opposite message to your body – that you are very safe. Nothing down-regulates the production of the alarm signals within your body more powerfully, as you could never breathe in this way if your life truly was in danger. No practice restores your energy and vitality like regularly breathing this way. And I don't say that lightly.

When you breathe in this way your tummy will move in and out, as opposed to just your upper chest. To try it, inhale slowly through your nostrils, allowing the lower part of your tummy to expand (imagining your tummy extending into the area where you can feel it meet your ribcage). Keep the slow inhalation going until your upper chest feels like it is pushing your ribs out at the sides of your body. Pause (don't 'hold' your breath) then allow the exhalation to begin in the reverse order of the inhalation, with the top and side of the chest emptying first, followed by the middle of your abdomen and lastly your tummy; your tummy shrinks back towards your spine. If that sounds or feels too tricky, just move your tummy (abdomen) in and out with each breath. Allow it to expand on the inhale and shrink back in on the exhale. It can help to place your hands on top of each other just below your navel. That way, you can focus on moving your hands in and out as you breathe. Be patient with yourself, for if you have been breathing in a short, sharp shallow way for months, years or even decades, then it can take some time before your abdomen wants to move in and out with your breath. Some people describe to me a feeling that they can't get their breath past their heart. This is often the case if people have felt anxious for quite a while. So be kind to yourself; in time and with practice the parts of you that have become disconnected will be thrilled to be back in touch.

At first, you will more than likely need to schedule regular diaphragmatic breathing time slots into your day, until it becomes your new way of breathing. Make appointments with yourself to breathe in this way. If it is peaceful each morning while you boil the kettle for the first time that day (to make your hot water with lemon, of course), instead of racing around and doing 85 jobs while the kettle boils, stand in your kitchen and breathe. Link breathing well to your daily routine, such as having a shower, or to a particular hour of the day, so that it quickly becomes a habit. Do it numerous times over the course of your day. Book a 'meeting with your breath' into your calendar each afternoon at 3 pm. If you work at a computer, have it pop up on the screen that it is time for your meeting with yourself to do 20 long, slow breaths. We keep appointments with other people, so be sure to keep the appointments you make with yourself.

‘Let there be room for not knowing. Some of the most beautiful chapters in our lives won’t have a title until much later. Don’t let the uncertainty deplete you. Embrace it and let the not knowing enliven you.’

FOCUS ON WHOLEFOODS

In chapter one I explained how what we eat affects our energy levels and in this chapter we've seen how stress can mess with our appetite, metabolism and weight, and increase our cravings for sugar. Many people know they need to eat less/no sugar, yet they don't do this. There are a wide variety of reasons for this, both biochemical and emotional. Over decades of working with people, I've met those who find taking a break from certain foods easy, while for others when it comes to food, they are 'more' people. For the latter group, instead of focusing on eating less sugar, they do better with their focus on eating more real wholefood, more vegetables and more dietary fats (from wholefoods). Increasing your intake of green vegetables and/or dietary fats from wholefood sources tends to make a big difference to your desire for sugar. A high intake of green leafy vegetables for a minimum of 21 days starts to change your taste preferences, as greens have a bitter taste.

When it comes to fat, if you have lived through the 'low-fat, high-carb' era and become overly anxious about your dietary fat intake, you may not be eating enough of it (see page 17). Notice when you crave sugar and significantly increase your intake of fats at the meal prior to the typical craving time. If the middle of the afternoon is your tough time – craving-wise and probably also with your energy – then eat more organic wholefood fats at lunchtime. Try adding plenty of avocado, nuts, seeds, olives, butter or coconut to see if these strategies work for you.

Even though fats have the highest number of calories per gram of the macronutrients, keep in the front of your mind that they will make you feel fuller for longer. Also, remember that carbohydrates lead the body to make insulin, which signals the body to store fat, whereas eating fats does not trigger fat-storage-signalling hormones.

DECREASE OR OMIT CAFFEINE

One of the first areas I encourage people to get honest about is how much caffeine they are having and how it is affecting them; most people have no idea about the latter. People tend to perceive that they 'need' caffeine for energy, yet years of working in health retreats where there was no caffeine on offer showed me week after week that this is not the case. In fact, hundreds of people would exclaim how much more energy they had after a week without it. I talked earlier about how caffeine activates our stress hormones (see page 50), so there's no need to repeat myself here, except to ask you to decrease it or omit it for a four-week trial. Some people do get headaches when they reduce or stop it, although interestingly this isn't dependent on how much they have. Keeping hydrated, as well as additional magnesium, can help alleviate the withdrawal headaches. If you notice that you do feel much calmer, then it's a signal to keep reducing your caffeine. Another option is to switch from coffee to green tea so that you consume smaller amounts of caffeine buffered by the effects of theanine in the green tea (see page 51 for more on tea). Please hear this: I am not anti-coffee or anti-caffeine. I simply want you to know that if you are stressed or feel anxious, adrenalin is one of the main hormones behind that. Consuming caffeine leads you to make more of this hormone. So why would you add to that?

CLOSE OPEN LOOPS

What else depletes energy? What I have come to call 'open loops'. These are the tasks we set ourselves every day but don't get around to completing (the emails, the calls, the household chores), so they float about in our heads, demanding to be dealt with, draining us of energy. It's like having a dozen programs running at the same time on your laptop – it drains the battery and slows everything down. Even worse, you may begin to judge yourself for not being able to close some of these loops. (Psychologists call this the Zeigarnik effect, where unfinished tasks contribute to intrusive, uncontrolled thoughts.) You may unconsciously berate yourself for not trying hard enough: 'Everyone else seems to be getting so much more done – they seem so much more organised, efficient, happy and successful.' This further depletes your energy, making it harder to take action.

The first step is accepting where you are right now. When you fight with what is, when you battle with the ways things are, you suffer. You are not lazy, hopeless or a loser. You are a person with a few things you'd like to get done, and you're going to start now.

1. Identify and capture in writing every open loop – the tasks you are thinking about that need resolution.
2. Organise these into simple headings that relate to the importance of each task and its urgency (priorities).
3. Schedule times to consider, reflect, decide and execute each task (see the next section). If you can, delegate some of your tasks (e.g. employ a cleaner, gardener or personal assistant), or ask for help from friends or family.
4. If it is not time to perform step 3 on the task or idea yet, then just know you have captured it and the time will come for you to schedule it.

TRY SCHEDULING EVERYTHING

Many people use to-do lists to try to manage their time, which may be helpful to some degree. However, to keep themselves motivated, most people tend to tackle those that can be completed quickly and may not even get to the tasks that they really need to do, such as exercising or having medical tests. As we saw in the previous section, another problem with to-do lists is that they can make people feel so overwhelmed that they don't know where to start.

Numerous research groups who examine time management and productivity have found that highly productive people do not work from a to-do list – they live and work from their calendars. If you are an ultra-busy person with a huge number of tasks that need completing in a day and/or you have other people relying on you, the only way to pull it all off is to prioritise and keep a schedule. If it doesn't get scheduled, it doesn't get done.

If you feel that your biggest energy-zapper is poor time management, then the following tips may help. Of course, these ideas won't be everyone's cup of tea. Scheduling every last thing might sound tedious, as if your life will lack spontaneity. But the truth is, it actually gives you the headspace for creativity and innovation to flow. Try it for a couple of days a week. You may find you get more done, and the stress will lessen as you have a sense that you are handling what's on your plate.

Scheduling tips

Pick a task duration. Most systems automatically schedule new events for 30 or 60 minutes' duration. Highly productive people only spend as much time as is necessary for each task.

Pre-schedule sacred time-blocks. First get clear on your personal priorities – things that align with your core life values. That might include meditation in the mornings, twice-weekly gym sessions, or a date night.

Now schedule work time-blocks. Time-management principles suggest that you schedule everything. Instead of checking emails every few minutes, schedule three times a day to do this. Instead of writing 'Call Sarah' on your to-do list, put it on your calendar or establish a recurring time-block each afternoon to 'Return phone calls'. And don't let your calendar fill up randomly by accepting every request that comes your way.

LEARN TO SAY NO

Clearly there are countless things we do or believe that can drain our energy. And saying yes when you really mean no is one of them.

Be honest with yourself. Are you someone who struggles to say no? Consider the energy drain on your body and mind when you agree to something when you'd prefer to be doing something else, or with other people, or by yourself, or catching up on some much-needed rest. And reflect on why you do it. Is it to please the person asking? But is it coming from a place of genuinely wanting to assist that person, or from a fear that they may not like you as much if you say no? The

reason matters. Because the more your choices are fear-based, the more exhausted you tend to feel. When your choices come from love, they typically energise you. Doing something out of duty depletes you. When you do something out of love, it energises you, and everyone around you.

If you struggle to say no, it is usually because you were raised to put other's needs before your own. And from this you will be seen as a very kind, caring, giving person. These words will be attached to your identity, which is why saying no is not as simple for you as it is for others. You have to see the benefit for the other person for you to say no, otherwise you never will. You have to see that in saying no, you give the other person the opportunity to grow, to find another way, to develop resources within themselves. The stress of obligation and the amount of time and headspace this takes up for so many people today is having a major impact on too many people's health and energy. See the gifts for others as well as yourself in your no. Live authentically.

STOP, KEEP, START

Another useful strategy is to take a journal or notepad and draw up three columns with the headings: 'Stop', 'Keep' and 'Start'. Now fill in each column as you think of things by answering the following questions:

- What am I going to stop doing?
- What am I going to keep doing?
- What am I going to start doing?

For example:

- Stop: *I am going to stop getting caught up in gossip as it is exhausting.*
- Keep: *I am going to keep eating a nourishing breakfast every day.*
- Start: *I am going to start walking four days out of seven, starting tomorrow at 6 am.*

Here are some other 'starts' that might inspire you:

- *I am going to start Stillness Through Movement classes.*
- *I am going to start lunchtime walks.*
- *I am going to take singing lessons.*
- *I am going to call/Skype my sister every week.*
- *I am going to start drinking warm lemon water first thing in the morning.*
- *I am going to start watching my children sleep (about an hour after they fall asleep) for 10 minutes each evening; they are little for such a short amount of time.*
- *I am going to start cleaning out my clothes cupboard every six months and give what I haven't worn in the last chunk of time to a charity store.*
- *I am going to stop responding to text messages immediately. This will give me a sense of freedom.*

CREATE A MORNING RITUAL

During my research for this book, I was fortunate to be able to interview some amazing people at a leadership retreat. Over the week, I asked some of the world's foremost minds and hearts about their own personal energy, how they experience energy, and what they do to cultivate energy. And one of the most common answers was to have a non-negotiable morning ritual that set them up for the day, physically and/or spiritually – rituals that often involved learning and growth. Many of these people, both men and women, had children ranging in age from 12 weeks to 21 years. Some had four children under 10. Those who did and who had a morning ritual of solitude either got up before everyone else in their household or, if they rose later, they had taught the children to occupy themselves while they did their yoga, meditation, reading, writing or other restorative or vigorous practice. Another key, however, was that they remained flexible so if a child needed them it didn't 'ruin' their morning. It was just what it was.

I want to encourage you to do the same. When you start your day with a meaningful ritual, it connects you to your essence and plugs you into your natural intelligence. Your mind is sharper, you hold your body more strongly and you are more effective in your work day. You cope better with whatever comes your way. Those I spoke to also suggested that they are better colleagues, parents, partners and friends when they have set up their day. So the ripple effect of a morning ritual is significant. Not only do you experience better and more consistent energy, but those around you get the best version of you, too.

BECOME AWARE OF YOUR LANGUAGE

The language you use, both habitually and sporadically, has a powerful impact on your energy. Picture this. You set your alarm for 6 am, planning to get up and exercise. When the alarm sounds at 6 am, your first thought is 'How can it be six already?' You're weary. So you press snooze. And you are still pressing snooze at 7 am when another family member wakes you fully from your slumber. Now being 7 am, there's no time to exercise as you have other responsibilities and then have to get to work.

Perhaps the language you use inside your head, having slept longer than you intended and having not done your exercise, is that you are 'lazy' or that you are a 'failure' or 'that's just typical', meaning that you already know you are a lazy failure and, well, you've just proven it again. What kind of energy does that language, that selection of words, and then the meanings they generate in your nervous system bring? Lousy energy.

However, perhaps your response to that scenario is to see it as it is, not better than it is and not worse than it is, just for how it actually is – you didn't get up and exercise. Full stop. If you see it in this latter way, your energy will be better, in part because the judgement you pass on yourself using the first set of language annihilates your nervous system. And it's the judgement – the way we berate ourselves and put ourselves down – that usually creates the same pattern the next morning. When you are telling yourself you are lazy and a failure, underneath it all you believe you are worthless. So why on earth would you take better care of yourself? And you just call it procrastination. And usually you keep judging. And that whole pattern is exhausting in itself.

Use full stops

When negative judgements are running in our head, most of us use commas in our inner dialogue where there needs to be a full stop. 'I didn't get up at 6 am and exercise this morning full stop' for many is 'I didn't get up at 6 am and exercise this morning, therefore I am lazy and if I am lazy, well that's what my father told me every time I didn't want to help him in the garden and if I'm lazy and he values hard work then he doesn't like or value me.' As crazy as that sounds, without realising it, that's what you can trigger when you judge yourself: a thought process like that. Not always, but pretty often. And that judgement is stifling and exhausting because it scratches your oldest wounds. The only thing is, you won't know what you have done to yourself. It all happens in the ANS – remember? The part of your nervous system that you can't access with your thoughts. It's all happened in a split second, without you literally giving it any conscious thought; you don't stand there in that situation and think 'How can I put myself in a really lousy headspace for the rest of the day?'

When you see things as they are, and you put the full stop where it is supposed to go, it fosters better energy because you don't scratch the itch that you are 'not enough'. For every human's greatest fear is that they are not enough, and if they are not enough, then they won't be loved.

Become aware of the language you use. It can make or break the energy of your day. Stand guard at the door of your mind and become conscious of how often you judge yourself. The second you notice it, employ the full-stop strategy.

> 'If you do something out of duty it will deplete you. If you do something out of love it will energise you.'

EXPLORE YOUR EMOTIONAL LANDSCAPE

Your behaviour is the outermost expression of your beliefs. The challenge is there isn't a subject at school that teaches us how to explore these and change them if that will serve our health and quality of life better. Plus, most of us absorbed many of the beliefs that create our behaviours before we were old enough to talk. And unless this is something you have explored in your adult life, they will be creating the life you now have. Some beliefs serve us. Others limit us, and we outgrow many of them, yet they still run our lives. The other tricky thing is that most of the beliefs that drive choices and behaviour are subconscious and, to understand just how important our beliefs are to our biology, we need to explore the subconscious as well as the conscious mind.

The conscious mind can think freely and create new ideas, and we are aware of this. However, the subconscious mind is more powerful and tends to be responsible for our pre-programmed behaviours, most of which we acquire before we reach the age of seven. It has been estimated that most of our decisions, actions, emotions and behaviour depend on the 95 per cent of brain activity that is beyond our conscious awareness, which means that 95–99 per cent of our lives comes from the programming in our subconscious minds.

The subconscious mind struggles to move outside its fixed 'ideas'. It also automatically reacts to situations with previously gathered observations to behavioural responses. This is often why we can be unaware of our behaviour; in fact, most of the time we are not even aware we are acting from our subconscious. Studies show that our brains begin to prepare for action just over a third of a second before we actually consciously decide to act. In other words, even when we 'think' we are conscious, it is our subconscious mind making the decision for us.

Your attitudes, beliefs and perceptions – positive or negative – can actually influence the health, behaviour and structure of the cells in your body. Becoming aware of them means accessing the behavioural patterns in your subconscious mind so that you can change the underlying limiting or self-sabotaging beliefs. Ones I often hear are 'I'm not good enough', 'I'm worthless',

A TYPICAL STRESS PATTERN IN ADULTHOOD

Something you weren't expecting happens: you lose your job, a relationship suddenly ends and you didn't want it to, someone you love passes away. Adrenalin production kicks in. You feel like you are in shock. You feel like you get more done – because you have to – and you sleep less. But you're not noticeably tired. If you stopped you would be; in fact, you worry that if you stopped you might never get up. But you can't stop, so you don't. You might also lose weight without trying. If the stress or the ramifications of the stress continue, your body can't sustain the output of adrenalin and all of its biochemical consequences (for example, inflammation). With the continuing stress your body has no choice but to start to make more cortisol than before, in an effort to dampen down the inflammation created by the ongoing, excess adrenalin.

You feel heavy in your body, perhaps lethargic, or just not as good as before when the stress was fresh. You may gain weight, a little or a lot. It doesn't seem to matter how little you eat or how often you exercise, nothing changes. You have thickened up through the middle, and it won't seem to budge. With the ongoing stress, you tend to drink alcohol each evening to avoid feeling the emotional pain, and so you wake up tired and start every day with coffee to get yourself going – two liver-loaders. This adds to the stress inside your body, takes away from the efficient production of energy, and over the long term can disrupt sex hormone balance and lead to fat accumulation in the liver, with energy continuing to decline and body fat continuing to increase.

No diet will resolve this. Resolving the emotional pain will. Resolving why you have stopped seeing and treating yourself as the precious human you are will.

'I would like myself more if I lost weight', 'I'm such a failure' . . . and the list goes on. You can see how incredibly destructive these beliefs are to your life – why would you make better-quality food choices when you believe you are worthless? You'll always find it an uphill battle until you address the (false) belief. If it is difficult for you to identify your core beliefs, it may help to ask a close, wise friend. Often they can see things that we can't. Or find an experienced psychologist who can help you explore your patterns.

The fatigue of failure

Here is a common example I have heard time and time again. You feel that you weren't listened to as a child and you weren't valued. Because of this you developed the belief you are worthless, not valued and not worthy of love. So now, as an adult, when you feel that someone is not listening to you, your biggest wound is reopened (remembering that our greatest fear is that we are not enough and if we are not enough then we won't be loved). It is hard-wired into our nervous system that love is essential for our survival, as for a human baby it is. Someone has to care enough about us to meet our basic needs of food, clothing and shelter. Other animals can survive on their own from birth. As adults, we know that a life with love in it is delicious but that we can *physically* survive without it.

Unless you have examined and chosen beliefs that serve you, you will usually respond from fear when this 'wound' is scratched and, before you know it, you've created a meaning that you are a failure, or that you let people down.

Deep down you feel hurt and vulnerable, as you did when you were a child, but to avoid feeling that vulnerability, you might explode in anger temporarily, creating a wake of emotional destruction around you. Anger is just sad's bodyguard though. And when you witness that destruction you judge yourself again – you're a failure as a mother/father, partner, friend, colleague – and you withdraw in a depressed mood, with the lowest energy you can imagine. Yet 10 minutes beforehand you'd had so much (albeit somewhat destructive) energy. So what has occurred here? It's what I call a 'crazy eight', which I represent with the infinity symbol: ∞.

At the left end is low energy, depressed mood, shallow breathing, slumped posture – a focus on the past. At the right end is the explosion: the anger, the forceful energy, the expression of pain. And at the midpoint is happy, contented, functional behaviour.

People who run crazy eights tend to spend very little time at this midpoint. They can spend days, weeks, months or longer in the low-energy state, passing constant negative judgements on themselves, telling themselves that they are a failure moment to moment. But no one can stay in that place forever. So to feel more, to have more energy, to pull themselves out of the low state, they explode, instead of seeing the emotional trigger for what it is: an opportunity to learn, grow and transform the wounds of the past and leave the crazy eight behind for good.

People who experience this tell me that the energy of the explosion feels cathartic but that they hate themselves for their behaviour. They believe that how they behaved in that moment is who they are – 'a failure', 'a bad person', unlovable. If your fatigue is the result of the meanings you create about who you are, the work you need to do to recover your energy is emotional.

Turning it around

As adults, I believe we not only have a responsibility to take care of our physical health but also our emotional health. In my experience, clients who are deeply fatigued (I'm not talking about physical illness or disease here) do not improve their energy levels until the meanings that were established in childhood are seen for what they are: a set of beliefs they developed as a way of surviving their environment as a very young and precious being. They will have moments or days when they feel a bit better, but the heaviness won't lift until that internal tug of war, the one that starts every morning about three seconds after they open their eyes, is resolved.

I believe that every emotion is here to serve us, and that what we feel gives us an opportunity to shine a light on something that may help us to grow or learn. Even when we meet real tragedy in life, we can react either by losing hope and falling into self-destructive habits, or by using the challenge to find our inner strength. That can mean sitting with physical pain, emotional pain, loss, heartache, heartbreak, disappointment, broken trust, anxieties or fears, and allowing ourselves to feel. Because what you feel gives you the opportunity to become of even greater service to the world. Quite often it is a case of the greater the wound, the bigger the gift. When you are able to see a 'purpose' to emotional pain, it is easier to bear and growth can result. Many adults, however, don't have the strategies to unravel emotional pain themselves. This is where the support of a psychologist or other health professional can help.

HOW 'KEEPING THE PEACE' DECLARES WAR ON YOUR HEALTH

Many people have shared with me that they keep the peace to avoid stress. Yet there is no peace when you have to keep the peace. Think about that. Here is a common example to help you see how everyday life, because of the perceptions we bring to it, can lead to cortisol, from an emotional source, being a contributor to health challenges.

Basic psychology teaches us that humans will do more to avoid pain than they will ever do to experience pleasure. Some women, in particular, will do anything to keep the peace and avoid conflict. Inwardly, they become highly strung because they are always walking on eggshells around others, especially their intimate partner, doing all they can to help prevent those around them from losing their temper. If the man of the house has a tendency to communicate with explosive, angry outbursts that seem unpredictable – hello, silent stress-hormone production! Some people avoid feeling emotional pain by eating too much or making other poor food or lifestyle choices – perhaps drinking bucket-loads of wine or chain-smoking cigarettes. Some go shopping and rack up credit-card bills that will take months or years to pay off. Alternatively, some people might cope or explore their pain by writing in a journal, or going for a walk or a swim. Others will pray, meditate, or telephone a friend and chat to deal with emotional pain.

Some of the ways we cope with what is going on support our health. Some potentially harm our health. And all of these activities may take place with or without a conscious understanding of why. I want to help you see why, so that you can change your response, if it is hurting your health, especially if the stress-hormone production triggered by your subconscious emotional responses is blocking you from experiencing the amazing human that you are – and is making you tired in the process. I love this quote by Marcus Aurelius, which may assist you if guilt or worry are patterns for you: 'If you are troubled by external circumstances, it is not the circumstances that trouble you, but your own perception of them – and they are in your power to change at any time.' Do something creative with that quote – stick it on your fridge if it helps, as a gentle reminder of a powerful truth.

The energy of appreciation

What you focus on is what you feel. In any given moment, if you are focusing on what is missing from your life, what is wrong with your life, and everything you wish you could change, you will notice that it does not foster great energy. In fact, it usually has the opposite effect – it is exhausting, which deprives you of the energy you'd need to make the changes you wish for anyway. But the minute you shift your focus to what you do have – what is already present in your life – you shift your chemistry, you shift your energy and you open your mind up to new possibilities.

It is not physically possible for the human brain to focus on two things at once. No matter how good you believe you are at multi-tasking, your brain can only focus on one thing at a time. And when you focus on what's good, what you appreciate and what you are grateful for, and you do this consistently, your energy shifts significantly.

Changing your focus to what you already have doesn't mean that you are wearing rose-coloured glasses or burying your head in the sand, or avoiding or denying what you would like to change in your life. You will still be aware that stuff is there. It's just that when you always focus on what's wrong, you don't have the energy to change it. When you focus on what you appreciate, it gives you the energy to start to change.

I witnessed this very powerfully in a lady who had been diagnosed with depression and was taking three different medications for it. Her energy was low, she was prone to constipation – typically, she used her bowels only once a week – and just sitting with her, you could feel her sadness. However, on another occasion I witnessed her (not in my clinic, but at an event) win a major prize and she jumped and squealed and smiled the biggest smile. It was delightful to watch, and I was so happy for her. In that moment her focus was elsewhere and her energy was far superior to what she had come to believe was normal for her.

The biggest obstacle to shifting our focus is that our emotions capture and blind us – anger, rage, frustration, disappointment, sadness . . . The more aware we become to the blinding capacity of our emotions, the more we realise that when we start getting angry or denigrating ourselves or craving things we don't have in a way that leads us to feel miserable, we begin to shut down. It is like sitting on top of a mountain with the most exquisite view laid out before your eyes, and you have wrapped yourself in blackout curtains.

You can experiment with this. You can go and sit where there is an amazing view. And the first hit is usually 'wow!' and your mind opens. Yet if you sit there long enough, you will start to worry about something. Then you realise that it feels as though everything is closing in and getting very small. We all do this, all day, every day. The skill is to catch ourselves and come back to the view, to the beauty, open to the big picture. Do this moment after moment after moment.

Which do you think fosters greater energy? The opening or the closing? The 'wow' or the worry? I'm not denying there aren't things to be concerned about; I am simply wanting to offer you a new perspective that encourages you to be open to wonder. To the wonder and gift of life, with all of its messiness and chaos and unpredictability. This way you actually get to enjoy your life instead of feeling that it is all too hard. You experience a spacious feeling about life, a calm and grateful heart – even if what you have to do in a day doesn't actually change. You enjoy your life and you are lit-up living.

When you fight with what is – the fact that you have a mortgage and bills and a demanding job, and challenges with your children and family, and people who need you – when you resist what is, you create a stress so deep within you that you no longer recognise the attitude you show up with every day. And how you have become saddens you. And the never-ending stress utterly exhausts you.

THE BLINDING NATURE OF GUILT

You may have created a life where all of your basic needs are met, and then on top of that you have a full-time job, you are doing some continuing education, you have three children and a partner, a house that needs cleaning, planes you have to catch on time, ageing parents who need your help ... and that is just the tip of the iceberg of your responsibilities and tasks. In such a space, if you cannot see the gift and the immense beauty in all of that – even amidst the chaos – you miss out on the extraordinary life you have created. Some can't see it this way and it devastates them. For others, it elicits major feelings of guilt. Trouble is, although there are people worse off than you in this world, the minute you feel guilty you change your focus so that you don't ever get the opportunity to identify what is bothering you, and more importantly why. What is bothering you can offer an insightful road to the core of something bigger, if you follow it with curiosity, rather than judging yourself. Constantly feeling guilty can be exhausting in itself. We think guilt is our conscience rising up to guide us. Yet guilt is really an interruption to the love that you are giving, sparked by scratching the itch of your belief that you cannot let anyone down. And what's behind that is age-old (false, unnecessary) fear that you aren't enough the way you are and therefore won't be loved.

The energy of purpose

Some people are low on energy because they believe they have to find their purpose to live fully. And although people with a purpose, with a mission to make a difference no matter how big or small, say this fuels their fire and gives them energy, those who don't have one, or haven't identified it, don't need to suffer. Because worrying about not having a purpose can be draining in itself. What people perceive to be social pressure to find their purpose, find their passion and know what it is that they want to do can have its own impact on energy. It is perfectly fine – and in fact recommended – to simply live each of your moments fully and marvel at it all. What if *that* is your purpose? Many small and large passions and purposes may enter and leave your life, all of them awakening something within you, allowing you to grow. For many people, I don't think there is some elusive purpose out there. It's here, within them, because they have senses to experience all of this. There doesn't have to be a big realisation, or bliss to follow, or a major moment of discovery of your life's purpose. Sure, for some this is their path, but for those for whom that isn't their way, it isn't sad, it's just what is. So maybe for a while, stop trying to find the forest and just enjoy the trees. And observe what that does to your energy.

However, that might not work for you. Maybe you still feel lost or as though something is missing in your life, or perhaps you want to make a bigger difference in the world than you feel you are now doing, but you aren't sure how. So many of us walk through life feeling numb and desperate for a deeper connection, but are not sure how to get it. Yet it's always in the giving.

Maybe the problem is not that you don't know what your purpose is. Maybe it is the *way* you are trying to find your purpose. We can't 'think' our way into our life's passion and purpose. We have to 'do' our way in. This means taking steps towards what you want and removing the things in your life that you don't want. And that might require minor or major changes that allow you to take one step at a time and explore many different passions. If you are searching for your purpose and some passion, try the strategy, 'Stop looking and start doing'.

The more we act, the more we get clear on things. So instead of over-thinking 'Will this work out?', 'Should I try that?', 'What if I don't like it?', start taking steps towards being able to try new things. The experience is the reward; clarity comes through the process of exploring.

It can also be a good idea to break up with the idea of 'the one'. Many of us struggle because we try to find that one thing that we are meant to do, but trying to find only one thing can be the reason why we feel like something is missing. The notion that we have only one thing we are meant for limits us from fulfilling our greatness. So start getting in touch with your passions and what you care about. When you lead a passionate life, you are living your life on purpose.

Consider that the real purpose of anyone's life is to be fully involved in living. Be present for the journey. Act on what you care about. If there are things in the world that are unacceptable to you, create a plan of action and take steps to make a difference in that area. While you focus on yourself, you will never find purpose. It has to be about others or there is no lit-up feeling inside. No matter how big or small, what is it that you really care about?

JOY

On my travels I met a lady who worked in a kitchen in one of the most beautiful places on earth. The view out across the ocean from the kitchen was stunning. When I remarked on this to her, she replied that she never sees it anymore, although she looks at it every day. She said she looks 'straight through the beauty'. At that moment, another staff member piped up and said, 'I suppose you only get to keep what you let yourself have.' At the time, I had no idea what he meant, but later that day, the sentiment he had offered hit me.

If it is right in front of you every day, and you don't let yourself have it – which means noticing it, taking it in, allowing yourself the pleasure of it – it is never yours. It never becomes part of the ground of goodness on which you can live, and this, the essence of this, forms so much of our foundation of energy that isn't really discussed. Because if you don't 'let yourself have' a spectacular view, then what else are you denying yourself? Why not let yourself have what you already have?

I have read that when people who are dying are asked what they will miss the most, they say, 'The ordinary things. The smell of air. The feeling of my dog's fur under my hands. My partner's face. A freshly cut lemon. The night sky.' We have those now. Let yourself have what you already have. It's what joy is all about. And joy gives us an irreplaceable depth of energy.

Signs that your stress levels need attention:

- you feel stressed regularly, like you are on 'red alert'
- you crave sugars and/or starches (carbohydrates)
- you love coffee and energy drinks, although sometimes they make your heart race
- you startle easily
- you don't sleep well on a regular basis
- you don't wake up restored or with good energy
- if you don't go to sleep by 10 pm, you get a second wind and end up staying awake until at least 1 am
- you sometimes wake up feeling like you've been hit by a bus – you are that tired
- you feel better if you can sleep until 8 or 9 am, rather than rising between 5 and 7 am
- you regularly feel tired but wired
- you are a worrier or a drama queen (or king)
- you feel anxious easily
- your breathing tends to be shallow and quite fast
- you experience 'air hunger' or shortness of breath (and other causes have been ruled out)
- you struggle to say no
- you laugh less than you used to
- you feel like everything is urgent and that there aren't enough hours in the day
- you notice that stress impacts your weight – either up or down
- you have put on weight around your middle and on the back of your arms, and you have grown what I lovingly refer to as a 'back verandah'
- you retain fluid
- your face looks puffy or swollen at times (and other causes have been ruled out)
- your body feels heavy and achy at times even though you don't have a medical condition that warrants this
- your blood pressure is high, low or the low end of normal
- you get dizzy easily, but particularly when you go from sitting to standing quickly.

Habits to better manage your stress levels

- One of the most important ways to manage stress is to practise conscious breathing. Diaphragmatic breathing (making your tummy move in and out as you breathe, as opposed to your upper chest) can literally change your life.
- Practise restorative yoga, pilates, tai chi or qi gong a minimum of twice a week for four weeks. Develop a daily practice for outstanding results. Commit to a breath-focused practice. My favourite is Stillness Through Movement.
- Spend five minutes daily focusing on all the aspects of your life for which you are grateful; you can't be stressed when you feel grateful.
- With the guidance of a herbalist, take some adrenal-supporting herbs. Not all herbal medicine is created equal, however, so you want to ensure the brand you take is reputable. Check that it has been tested by a high-level quality-control system, it contains no contaminants and that the active ingredients said to be in the product are indeed present. Herbal medicine can be taken in a liquid tincture form or as tablets. The majority of the following adrenal herbs are adaptogens, meaning they help the body adapt to stress by fine-tuning the stress response. They include: withania, rhodiola, Siberian ginseng, panex ginseng, licorice (if your blood pressure is on the low end of normal), dandelion leaves (if you retain fluid). The adrenals also love vitamins B and C and magnesium.

The restorative power of good food

Although I often recommend supplements of herbs and/or nutrients for adrenal fatigue, never underestimate the healing and restorative power of food the way it comes in nature. Taking supplements is not a reason to eat a poor-quality, low-nutrient diet. I simply recommend supplements where appropriate and to assist in the restoration of health. There is no pill that can make up for a lousy way of eating and how you nourish yourself with food is the basis of all self-care and energy.

CHAPTER FIVE

HEALTH CHALLENGES & ENERGY LEVELS

In this chapter I look at some of the health challenges that lead to fatigue. In most cases, these will require the support and guidance of a health professional, but I include them here to open your mind to the myriad roads that can lead to exhaustion. Understanding what is creating your poor energy is *essential,* as correcting what has gone awry in your individual biochemical, nutritional and/or emotional landscape depends on knowing what created it.

POST-VIRAL FATIGUE

Fatigue can follow any infective event, and in practice I have seen it regularly after an upper respiratory tract infection, usually caused by a *Streptococcus* species. I've seen it after Q fever, Ross River fever and what is considered to be a form of Lyme disease. (As an aside, I acknowledge that Lyme disease, caused by the *Borrelia* species, is not currently recognised as being in Australia; however, I do believe that an infective organism of a similar nature to *Borrelia* will be identified and found to cause symptoms identical to, or similar to, those of Lyme disease. In my clinical experience, those who have been diagnosed as having Lyme disease are often the most unwell and need urgent support for many body systems.)

Fatigue can also be a symptom following what I call 'Bali belly', where someone has had food poisoning or picked up a bug from water and, after the diarrhoea has resolved, they are left with what is diagnosed by their GP or a gastroenterologist as irritable bowel syndrome (IBS). Some people experience a debilitating fatigue along with their gut symptoms. Many also describe a brain fog or thick, heavy feeling in their head. Given that many neurotransmitters, such as serotonin, are predominantly produced in the gut and have most of their receptors in the gut, this is not surprising.

Glandular fever (also known as mononucleosis) is an infection caused by the Epstein–Barr virus and is well-known to cause fatigue. Some people, in fact, experience a fatigue so deep that they still don't feel like they are over it, even though the infection itself may have occurred more than a decade ago. Yet there are others who get this virus and don't even know they've had it.

Infection is clearly another pathway to poor energy, as the infective organism(s) divert resources for their own use and chronically activate the immune system, alerting many processes inside us. Therefore, supporting the immune system must be a priority for optimal energy and health. Sometimes when I talk about immune function, I sense that the importance of this system doesn't fully register with some people. Many people simply link great immune function with minimising how many colds and flus they get. Yet it is not only critical for the prevention of viral infections, but also in the prevention of autoimmune diseases such as multiple sclerosis, lupus, Hashimoto's thyroiditis, Graves' disease and coeliac disease, all of which are on the rise. Taking great care of your immune system is of immense importance to your long-term health and quality of life. Since most of your immune system lines your gut, a robust immune system must begin with good digestion and optimal nutrition from real wholefoods. Sleep also plays an enormous role.

'Supporting the immune system must be a priority for great energy.'

Many of the people who end up with ongoing debilitating fatigue after a virus have typically also had massive family stress. Time and time again I have met clients who tell me that they experienced immense fear and/or rejection growing up, whether due to a parent having an addiction or mental illness, or simply their perception as a child (they may have been safe and loved, but they didn't feel that way). This is why a holistic approach to health is required. There are biochemical, nutritional and emotional components to people's wellbeing, and addressing all three (what I call my 'three-pillar approach') is usually critical to supporting them to become well again.

EPIGENETICS: THE EMPOWERING NEW SCIENCE

In my opinion, epigenetics is one of the most empowering movements in science today. It may sound futuristic and complex, but the key concept is really quite simple. Epigenetics means you have the ability to control the expression of your genes – in other words you have the ability to control to what extent you 'inherit' certain health conditions. Until fairly recently, science suggested that the genes you inherit from your parents – namely, the ones that code for disease – are the most common way to inherit risks for diseases like cancer and heart problems. But there is another way to pick up genetic changes that researchers are starting to pay closer attention to. And I love it, because it puts you back in the driver's seat of your health. Because, let's face it, when you believe something is genetic, do you feel like there is anything you can do about this? No. So you spend your life living in the cloud of false belief that you have to put up with your health picture, when there may be numerous options for you to employ to resolve a particular health challenge.

Most of us had our first taste of science and biology (whether we were awake or not!) at school. As part of the curriculum we learn that genes, made up of DNA, are essentially the blueprints that make us who we are. This DNA code we have is a unique combination of information from our mother and father. Which genes we pick up from mum and which from dad is a somewhat random process. That genetic draw in turn helps to determine which diseases we're most at risk of developing during our lifetimes – well at least that's how we used to view it.

More recently, scientists have learned that DNA alone does not define our health destiny, and hence they have been focusing on another layer of genetic inheritance or potential: epigenetics. Epigenetics also plays an important role in determining what our DNA blueprints become. Epigenetic changes can be created by exposure to things like smoking, environmental pollutants and nutritional factors, as well as lifestyle behaviours. Essentially our environment has the ability to switch our genes on or off, which of course has significant implications when it comes to turning on or off the genes relating to disease. Furthermore, researchers have demonstrated how it is possible to pass on these epigenetic changes to your children. So how you eat, drink, move, sleep, think, breathe, believe and perceive literally has the ability to influence your genetic blueprint. It is truly mind-blowing.

This is particularly exciting as we hear of people likely to have inherited disease risk factors feeling completely disempowered by their luck of the draw. We can move from feeling 'trapped' or 'defined' by our genes to being empowered that the lifestyle we lead can literally alter the expression of these genes. The power is well and truly within our hands – or perhaps more accurately, on the end of our forks.

HISTAMINE & FATIGUE

If people experience allergic symptoms, such as sneezing, a blocked or runny nose, sinus problems, itchy eyes or hay fever, they can feel fatigued after a particularly reactive day. Or it may be a time of year that is worse for them, such as spring. One of the substances involved in an allergic reaction is histamine, which is released from a type of immune system cell, called a mast cell, when its cell wall becomes unstable. The best way to picture this is as a circle and the outside line of the circle (the cell wall) starts to vibrate when it is unstable. When this occurs histamine is released, and this is one mechanism through which allergic symptoms can start to present. The level of exhaustion that this alone can cause for an individual is significant, and so it is important to address when someone is seeking improvements in their energy.

There are many aspects to histamine being released. For example, mast cell walls require vitamin C for their stability, so if someone has a poor dietary intake of this vitamin, histamine release is more likely. As an aside, many people with hay-fever type symptoms do well with daily supplements of vitamin C at larger doses (2–5 grams).

Another reason histamine reactions, and hence fatigue, can become more likely is due to poor adrenal function. Not only is histamine released in many allergic reactions, but other substances that drive inflammation are as well.

Inflammation can be created from a wide variety of scenarios in the body, one of which is long-term stress driven by adrenalin. Being the acute (short-term) stress hormone, adrenalin is designed to be produced for only very brief periods of time and in short bursts to get us out of danger. But we have seen that adrenalin production in modern times can be constant and relentless, for a variety of reasons. The body cannot sustain the metabolic consequences of these consistently high levels of adrenalin output, and so the adrenals release more cortisol to dampen down the resulting inflammation. Elevated cortisol comes with its own problems and also requires increased amounts of vitamin C to produce it. So if someone has a poor dietary intake, or their lifestyle induces the destruction of vitamin C – such as smoking or being exposed to other pollutants – the adrenals get to a point where they can no longer sustain the output of cortisol required to keep the inflammation at bay.

The individual then ends up with adrenal fatigue, and the subsequent deep, unrelenting fatigue this creates. When cortisol levels have fallen low, the ability to counteract the inflammatory, allergic-type reactions becomes poor and histamine release tends to increase, driving the allergic symptoms. People experiencing adrenal fatigue often notice that they become more sensitive or their allergies get worse when their cortisol is particularly low.

Conversely, the more histamine that is released, the harder the adrenals have to work to try to produce enough cortisol to dampen down the inflammation, and the more worn-out the adrenal glands themselves appear to become. This can establish a vicious cycle of reduced cortisol allowing histamine to inflame the tissues, leading to worsening adrenal fatigue, as well as more extreme allergic responses. And the fatigue that goes along with all of this can be debilitating.

Actively supporting your adrenal glands and the immune response through the use of specific nutrients such as vitamin C and bioflavonoids, and herbal medicine, as well as eliminating or reducing your exposure to foods and other substances that cause allergic or sensitivity reactions, can help break this cycle as well as strengthen adrenal function. Problematic foods for an individual can interfere with daily functioning and become a profound stress on the adrenal glands, so it is important to work with an experienced health professional to determine and trial an elimination of these foods to both decrease the load that the food reactions can bring, and to promote adrenal health. Dietary changes are best guided by a nutrition professional.

ADRENAL FATIGUE

If you have had a high level of cortisol output for many, many years, your adrenal glands may not be able to sustain this. They are not designed to withstand this kind of output, and so they 'crash', and cortisol levels plummet.

Our bodies cope with this long-term stress in different ways. For some, adrenalin seems to remain as the dominant stress hormone, while others flip over into a more cortisol-dominant stress response. If the stress response doesn't truly switch off, there is the potential that the adrenals will eventually run out of substances necessary to create the hormones they are being instructed to produce (by the pituitary, based on its perception that your life is in danger), and cortisol output will no longer be optimum or elevated; it will be negligible. This has become known as adrenal fatigue, because the major symptom is a deep, unrelenting fatigue. In general terms, you burn out; you feel like you've been hit by a bus most days and it is worst in the mornings.

However, at its extreme, this can become a condition called Addison's disease (although there can be antibodies involved in this condition). If a person's cortisol level is extremely low but still falls just inside the 'normal range', they will be told that they are fine. They may feel lousy, but all the tests they have always come back 'normal'. Yet they feel anything but normal, and people who know and love them will often comment that they are a shell of their former selves.

Cortisol can also be rather sinister in that it can interfere with your steroid (sex) hormone metabolism, sleep patterns, mood, and insulin and blood glucose management, as well as thyroid function: all elements that influence energy, metabolism and how you feel and function every day.

In my opinion, practising a form of restorative yoga, or Stillness Through Movement, combined with dietary changes and herbal medicine (see page 65 for more on this) are the most effective ways to help you recover from this debilitating condition.

LIVER DETOXIFICATION

The liver has a crucial role to play in every aspect of our health, including our hormone balance, fat metabolism, the clarity of our skin and eyes, and our vitality. In conjunction with the gallbladder, the liver works endlessly to help us excrete unwanted substances such as old hormones, pesticides and stored body fat. And if these substances are allowed to accumulate (as described in the toxicity section on page 75), energy can be significantly hampered.

Physiologically, the liver sits behind our right ribcage and its primary role is detoxification. Amazingly, it's the only visceral organ in the body that can regenerate new cells when old ones die. However, if we overload the liver, eventually it cannot keep up with its tasks and globules of fat can take up residence where working cells used to be. As more of the cells die, more fatty deposits will continue to accumulate – a condition known as Non-alcoholic Fatty Liver Disease. This is where our health and energy can really begin to suffer. Less efficient detoxification processes can lead to poor thyroid function, sex-hormone imbalances, congested skin, lousy cholesterol profiles and impaired blood-glucose management, which often shows up as sugar cravings. Moreover, *where* our body wants to lay down body fat can also shift. For the first time, people may notice that they have a fat roll quite high up on their abdomen. For women this is just below their bra line, and for men, just beneath their pectoral muscles. It can come and go, and sometimes there is a point right in the centre of the torso that is tender. I will always suggest ways to support the liver based on the presence of a fat roll in that position, and gallbladder support based on that tender point. Sometimes both organs need attention.

Detoxification is essentially a transformation process. Any substance that would be harmful to you if it accumulated in your body must be changed into a less harmful form so that it can then be excreted safely from your body. The choices you make influence how efficiently the liver is able to do its job, and this significantly contributes to how you feel. In addition to the liver, there are numerous organs and body systems involved in detoxification, including:

- The colon (digestive system) – you want to keep your bowel moving regularly, as one of its roles is to release waste and problematic substances so they don't accumulate (see page 24 for more on digestive health).
- The kidneys – these filter your blood and expel anything your body doesn't need in urine (see page 33 for more on kidney function).

- The skin – not only protects and houses your organs, but allows problematic substances to leave the body via perspiration.
- The respiratory system – lungs are responsible for filtering out fumes, allergens, mould and airborne toxins (even the hairs inside your nose help filter the air you inhale). When we are stressed, we tend to shift from slow belly-breathing to short, shallow upper-chest breaths, which in turn can reduce the lungs' ability to transport oxygen to all tissues. You now know another reason (other than the nervous system and stress-hormone-lowering benefits) why diaphragmatic breathing is my number-one health tip.

LIVER-LOADERS

Liver-loaders are what I call the substances that our liver needs to transform before they can be excreted. They can be roughly divided into two groups.

- Substances we consume or absorb through our skin:
  > alcohol
  > trans fats
  > refined sugars
  > caffeine
  > pathogens
  > synthetic chemicals in pesticides, skincare products, laundry and cleaning products and medications.
- Substances that our bodies make:
  > cholesterol
  > steroid (sex) hormones, such as oestrogen
  > metabolites created by or causing any short-fall in digestion, due to compromised digestive processes
  > those generated by untreated food sensitivities
  > metabolites produced in people eating gluten who have (undiagnosed) coeliac disease.

Alcohol

Unfortunately, alcohol is a depressant in its action on the nervous system. And in the aftermath of too much to drink, many people can identify with the fatigue that is present. Yet, if you wake up tired every morning and you tell me that you drink alcohol every day – even if it is only two glasses – I will tell you that this habit might very easily be the thing robbing you of your get-up-and-go.

Now, I don't say what I'm about to say lightly. There is a reason alcohol is at the top of the list of liver-loaders: it is quite simply a poison to the human body. It is a poison because we cannot excrete it. When we consume alcohol, the liver has to transform (detoxify) that alcohol into another substance called acetaldehyde, which is what we are then able to excrete. And our bodies will prioritise this detoxification, because if alcohol is allowed to accumulate in our blood supply to an exceptionally high level, we will go into a coma. That is how it works and we cannot fight our biology.

Yet too many people regularly over-consume alcohol, to the detriment of their health and energy levels. Whether it is increased body fat or cellulite, worse bouts of premenstrual syndrome (PMS), mood fluctuations, or perhaps your get-up-and-go has got up and left, the price of over-consumption is just not worth it. And that's just the effects on an individual level – alcohol also affects our relationships, our productivity at work and our safety on the road . . . the list is endless. As a society, we need to get real about the dangers of the regular over-consumption of alcohol – effects that are far-reaching.

So what is 'over-consumption'? The latest National Health and Medical Research Council guidelines recommend no more than two standard drinks per day, for both men and women, but that we must have at least two alcohol-free days per week. A standard drink is 10 grams of alcohol, in whatever form that comes. In Australia and New Zealand, 10 grams of alcohol is a 330 millilitre bottle of 4 per cent beer, a 30 millilitre nip of spirits, 170 millilitres of champagne, or a measly 100 millilitres of wine – about four swallows! Next time you pour yourself a glass of wine, measure it, and see what your natural pour is. For most, it is considerably more than 100 millilitres, and, as a result, many people unknowingly over-drink.

I would also encourage you to consider the position statement on alcohol endorsed by many of the cancer organisations from around the world, which says that if you have a family history of cancer, there is no safe level of alcohol consumption – which is a very powerful statement to contemplate. The link between the consistent over-consumption of alcohol and breast cancer is undeniable. Research has shown this time and time again and for many years now, yet we rarely hear about it.

I am not suggesting that you don't drink. Having a drink can be immensely pleasurable for those who partake on occasion. I simply want to appeal to you to get honest with yourself about how much you are drinking and how it may be affecting you. You know in your heart whether you drink too much and when it is negatively impacting your health. Alcohol can affect the way we relate to those we love most in the world, and how we feel about ourselves and, of course, deplete our energy.

If you drink too much, you must enquire within to examine if there is a pain (usually emotional) that you are trying to numb. Unless you also address the underlying reason you drink, it can be hard to cut down. Remember your behaviour is the expression of your beliefs so speaking to a counsellor or therapist may help you gain insight and new habits.

Synthetic chemicals

You only have to look at the way nicotine patches work to realise that the skin provides a direct route to our bloodstream. These products can contain substances that require change before they can be excreted.

- *Skincare products*
  There are plenty of wonderful skincare companies out there who do not use synthetic ingredients. Seek them out or even make your own. I love to suggest to people that it would be good if they could eat their skincare! You want to be able to recognise the words on the label of your skincare, just as you do with food.
- *Laundry and cleaning products*
  You also want to be able to recognise the ingredients in your laundry and cleaning products, as these easily enter the body through your skin or respiratory system. There are some wonderful, responsible companies out there making effective cleaning without synthetic chemicals simple – better for you and the planet. Do your own research on this one, and if in doubt, use vinegar and bicarb soda like our grandmothers did.
- *Pesticides and herbicides*
  It is also important to do what we can to minimise our exposure to, and consumption of, pesticides and herbicides. First, a number of these synthetic chemicals mimic oestrogen and can bind to the oestrogen receptors in the body, which has consequences for males and females of all ages. Research from the United States suggests that a significant number of girls are now starting to menstruate at the age of eight. It is difficult to explain how this is so without contemplating the role of environmental oestrogens.

  Another concern with the consumption of pesticides and herbicides is the risk of their storage in the fatty tissue of our body. We don't know the long-term consequences of this, nor of being exposed to these substances for an entire lifetime, as we are essentially the first generation of people to be exposed to some of them for such a long period.

OESTROGEN METABOLISM

To understand an additional mechanism of fatigue, oestrogen metabolism needs to be explored, a process relevant to men, women, boys and girls. When you have too much oestrogen – regardless of where it has come from – it doesn't just put you at risk of ill health, it can also be debilitating to energy. Oestrogen dominance can drive a bone-weary tiredness that is like no other. So let's understand its metabolism better.

Oestrogen can originate from inside the body or come from the environment – plastics, pesticides and foods may contain substances that can bind to oestrogen receptors in the body. In menstruating women, oestrogen is made by the ovaries. But for everyone else (men, boys, girls and pre-pubertal or post-menopausal women) oestrogen is made predominantly by the adrenal glands and the fat cells. In other words, the more fat cells we have and the bigger they are, the more oestrogen we make, regardless of our gender. This is a major problem for numerous aspects of our health, but also for our energy levels. When it comes to sex hormones, ratios are everything.

Like any liver-loader, when oestrogen arrives at the 'front door' of the liver, it undergoes its first stage of change (phase 1 liver detoxification). Between the front door and the 'middle' of the liver, oestrogen is still oestrogen, but it has been altered somewhat. This slightly changed form of oestrogen then enters one of the five phase 2 detox pathways to be altered again, and it is this substance that can then be excreted.

Health problems can arise, however, when the traffic on the phase 2 pathways gets banked up like traffic on a motorway. After years of regularly consumed liver-loaders, and/or hormonal or bowel problems, the roads out of the liver can become congested. Conventional blood tests for liver function do not reveal this. The liver usually takes years of battering before conventional blood tests reflect the congestion that led to the liver function enzymes becoming elevated in the first place. When the traffic is banked up, the oestrogen undergoes its first stage of change and arrives in the middle of the liver, ready to go down the appropriate path for the second stage of transformation before excretion. If the phase 2 pathways are congested, the oestrogen sitting in the middle of the liver has nowhere to go. However, it cannot remain waiting in the middle of the liver, as there is more rubbish constantly coming through the front door. When this occurs, the liver releases the oestrogen back out into the blood, and it gets recycled. It is this recycling of substances, rather than the substances themselves, that can potentially be harmful to human health, including our energy and our waistlines, due to the metabolic effects that the increased levels generate. In addition, it is this recycled form of oestrogen that has been linked to numerous reproductive cancers, particularly oestrogen-sensitive breast cancer.

What organ should we take much better care of if we want to stop this recycling from happening? Our precious liver. Our liver needs more support, and less of a load, for amazing energy, vitality, disease prevention and, hence, longevity.

TOXICITY, BODY FAT & ENERGY

As we saw in chapter one, body fat is a major source of fuel for our bodies. But it also has another function, as a storage house for what I call 'problematic substances', which your body might ingest or absorb. While their road into the body might vary, once they are in the blood the body has to deal with them, as problematic substances cannot be allowed to accumulate. So the body shunts them to the liver and the kidneys for detoxification, filtration and then excretion.

But here's the thing: if the load arriving at either of these organs is too much for it to deal with at the time, either due to a backlog or a lack of detoxification enzymes (specific nutrients are essential for the production of detoxification enzymes), the problematic substances still need to be removed from the blood so that they don't harm you. In such cases they are moved to the fatty tissues of your body (unfortunately some are stored in bones or the brain as well), away from the vital organs. The fat cells thus become storage houses for toxins – substances that the body sent to the liver to be changed so that they could be excreted, yet instead are recycled.

They were sent to the liver because your body deemed them unsuitable to stay. In fact, your body may have identified them as substances that could put your health at risk if they stayed. But because the liver didn't have the capacity to handle what arrived, the problematic substances were slightly (not completely) changed, becoming more reactive and problematic than they were in their original form that entered the liver. And once they ended up back out in your blood, your body knew they would be even more of a problem for you if they stayed there! So to get them (literally) out of circulation, they are sent to your body fat to be stored until some more resources (usually nutrients) are available to fully detoxify them. Only then can they depart your body for good. But most people are never taught this approach to their health, and as a result are never taught or supported to create the space (consume fewer liver-loaders) and enhance the resources (consume more nutrients) so that these problematic substances can be released from storage and excreted. Instead they stay there, and more and more body fat, and more and more fatigue, results. A poor state of mind can also unfold when this is going on.

To burn the stored fat as fuel and hence lose weight, you will also need to release these stored toxins, and your body can only do that when it has the required detoxification and filtration capacity. This means you need to put fewer problematic substances in, and enhance your nutritional status through increasing the nutrient density of your diet. Not eat fewer calories. Moreover, if the low-calorie foods come out of packets, they potentially contain some problematic substances, adding to your load.

If you don't create the space for your liver to have to do less detoxification work by decreasing the loaders going in, and you don't increase the nutrients you ingest to allow the liver to create more of the enzymes required for the critical detoxification work, then fat won't be burnt and energy won't be enhanced, as you are not freeing up the required detoxification capacity.

It is not a pleasant thing to consider, but essentially people are getting more and more toxic due to their lifestyle choices, and due to a lack of awareness about the chemical load going into their bodies via what they eat, drink, inhale from the air, or absorb through the skin. The more rubbish we store inside our bodies, the more lethargic we feel.

For some people, their inability to lose weight despite eating fewer calories than they burn – and particularly when the calories consumed are not from real, nutrient-dense wholefoods – can be due to these mechanisms. The body will still not have the resources to allow you to burn your body fat and enable you to lose weight. For to do so risks dumping the problematic substances back into your blood, which could harm you in a minor or major way. And your body is so clever it won't allow this to happen, so instead it continues to store the toxins until it can process them. And the lethargy that can stem from this is often debilitating.

LIVER HONESTY

At the weekend events that I run, I help people understand the exhaustion and guilt a vicious cycle of too much alcohol in the evenings, coffee to cope the next morning, and sugar cravings mid-afternoon drives. The guilt you feel tends to lead you back to that cycle, whereas if you saw it as a one-off and then returned to a higher level of self-care, it would have much less impact. Remember, it is what you do every day that impacts your health, not what you do sometimes.

Be honest with yourself about the liver-loaders in your life. Focus on taking good care of yourself and nourishing yourself, rather than on what you may need to consume less of, as that in itself can feel overwhelming or exhausting and just another thing to do. As you have read, the liver plays a significant role in our experience of energy as well as in the metabolism of countless substances that are linked to whether the body gets the message to burn body fat or store it, not to mention disease prevention. We only have one liver. Love it accordingly.

Signs that your liver needs support:

- you have a 'liver roll' – a roll of fat that sits high up under the bra line for women, under the sternum for men
- you have a tender point in the centre of your torso, which can indicate gall bladder issues; if your gall bladder has been removed, your liver has to make the bile on demand that the gall bladder once stored, so additional liver support is often required
- you have a very short temper and are easily irritated or impatient
- you experience episodes of intense anger
- you have pre-menstrual stress
- the presence of cellulite
- you overheat easily, or wake up hot during the night
- you have 'floaters' in your vision (and iron deficiency has been ruled out)
- you wake around 2 am
- you sleep poorly after drinking
- you're not hungry for breakfast when you first get up in the morning
- you prefer to start your day with coffee
- you have elevated blood cholesterol
- you bloat easily
- you drink alcohol daily.

Habits to better support your liver function

Set yourself a time-based goal when it comes to making these changes. For example, 'I will only drink alcohol on weekends for four weeks,' or 'I will only drink coffee when I go out for breakfast on Sundays.'

- Take a complete break from alcohol.
- Only drink alcohol on weekends.
- Replace coffee with green or white tea (or, less often, weak black tea).
- Support the liver with herbs such as St Mary's Thistle (especially if you drink alcohol regularly), globe artichoke (especially if you have a tendency to constipation, and/or a liver roll and/or central torso tenderness), turmeric (a brilliant anti-inflammatory), gentian (a wonderful bitter herb), bupleurum (especially if there are clots in your menstrual blood) and schisandra (for its detoxification action, plus it supports the adrenals).
- Drink vegetable juice or a 'green smoothie' most mornings, or start the day with an organic green powder – made from green vegetables – mixed with water.
- Omit processed food completely, or only eat it once a week.

- Cut down on eating fruit if you have more than two pieces per day, and don't consume any after morning tea.
- Take an essential fatty acid supplement.
- Eat high-zinc foods such as oysters (from clean waters), or take a zinc supplement of zinc picolinate (15–30 milligrams per day; best taken at night just before bed to maximise absorption).
- Transform anger into passion by giving a different meaning to a past experience (the energy of anger and passion are similar; they are just directed very differently).

THYROID FUNCTION

The thyroid gland is a little butterfly-shaped organ that sits in your throat and produces thyroid hormones. Thyroid hormones are like the spark-plugs of the body. They ignite the fuel in the mitochondria of each cell, which is necessary to produce the biochemical energy with which the body performs all of its functions. For this reason, any problem with the thyroid gland leads to widespread energy problems and often an unrelenting fatigue. Body fat levels can also change.

Thyroid hormones increase the metabolic rate, as well as speed up the rate of oxidation occurring in the body. (Remember, the oxidation process generates free radicals and antioxidants are required to stop the free radicals from damaging body tissues.) The metabolic rate, in turn, impacts on *every aspect of body functioning*. This includes the metabolism of fats, carbohydrates and proteins, digestion and cardiovascular health. It affects DNA and protein synthesis, body weight, heart rate, blood pressure, respiration, muscle strength, sleep, body temperature and sexual functioning, to name just a few processes. Thyroid hormones impact on every body system, and energy levels – among many other things – are severely compromised when levels fall too low.

A number of nutrients are essential to the production of optimal levels of thyroid hormones. Iodine and selenium are both vital trace minerals to this process of conversion, which helps generate an energised feeling and literally lights up your metabolic rate. Many people today get very little iodine and selenium in their diets, as the majority of soils in the Western world are very low in, or do not contain, these essential trace minerals. And if a nutrient is not present in the soil, it cannot be present in the food grown in it.

Iron is another mineral critical to the creation of healthy thyroid function, yet iron deficiency is the most common nutrient deficiency in the world (see page 22 for more on iron).

If iodine, selenium and/or iron are deficient, the conversion of inactive thyroid hormone into the active thyroid hormone won't happen efficiently and the thyroid can become underactive (hypothyroidism). However, nutritional deficiencies are not the only way thyroid function can be disrupted; long-term oestrogen dominance also interferes with good thyroid function.

So, the thyroid gland may not be working optimally due to any of a variety of causes – nutrient deficiencies, the over-consumption of substances that can interfere with optimal thyroid function, oestrogen dominance, or infection. Hence, knowing *why* someone's thyroid isn't working optimally is *essential* to its treatment. The road that created poor thyroid function is the same one that needs to be taken out of the problem. For example, if the thyroid isn't working properly due to iodine deficiency, then correcting the iodine deficiency will restore thyroid function. But if the hypothyroidism has been created by long-term oestrogen dominance, then no amount of iodine will correct that. Resolving the oestrogen dominance will. Road in equals road out.

THYROID DISEASES

A growing number of people in developed countries are experiencing thyroid problems. Some involve thyroid hormones only – T4 and T3, for example – the production of which are generated by a hormone made by the pituitary called thyroid stimulating hormone (TSH). Other thyroid conditions involve the production of antibodies, such as Hashimoto's thyroiditis (hypothyroid with antibodies) and Graves' disease (hyperthyroid with antibodies). With the latter condition, there is an increase in the metabolic rate (due to the thyroid being overactive), which can create an intense feeling of energy. (Some people describe it as a hyperactive or anxious feeling that they can't control.) Autoimmune diseases of the thyroid have increased significantly in the recent past. Current scientific evidence says that this is when your immune system, which is supposed to defend you from infection, starts to see your own body tissue – in this case the thyroid gland – as a foreign substance, like a germ, and attacks it, leading to a change in its function. This can lead to either the overactive picture (Graves' disease) or the underactive picture (Hashimoto's thyroiditis), a key symptom of which is a deep, unrelenting fatigue. As a side note, after doing my PhD and over two decades of working with people, I personally believe it will only be a matter of time before science determines that autoimmune conditions aren't due to a rogue immune system, but rather that the original infection is still there, albeit in a chronic, low-grade way. We just don't have the technology to find the pathogens yet.

If you have not been diagnosed with a thyroid illness, but you exhibit numerous symptoms, do not rely solely on your blood-test results to determine the diagnosis. It is beyond the scope of this book to explain why these tests may not tell the whole story (my other books explain this), so for now, work with a health professional who will treat the symptoms rather than rely solely on the bloods, and who will monitor both your symptoms and the blood work as you explore treatments. Let the blood work be a guide, not the be-all and end-all.

Please note that many of the symptoms of an overactive thyroid are often the opposite of an underactive thyroid; some people may experience both conditions in their lifetime. Given an underactive thyroid is more related to exhaustion than an overactive thyroid, this section focuses predominantly on the underactive state.

Signs of an underactive thyroid:

- you have experienced unexplained weight gain
- you feel cold in your bones, or are colder than others around you (you are the first to put on a jumper)
- you have a tendency to constipation, dry skin, brittle hair
- you suffer long-term oestrogen dominance symptoms, such as pre-menstrual stress
- you are weary to your bones – beyond tired – and your body feels heavy and lethargic
- your reactions to stimuli – both physical and emotional – feel slow
- you crave salt
- you crave coffee but it doesn't amp you up – your brain feels slightly more functional afterwards
- your groin aches
- your voice has changed; it is husky on occasions, particularly when you are extra tired (this can also be a sign that the adrenals need support)
- you feel like you retain fluid
- you often experience a depressed mood, forgetfulness and a sense of being easily confused
- you experience hair loss, or loss of hair of the outer third of the eyebrow
- you have difficulty conceiving
- you have challenges with menstruation
- you experience recurrent headaches
- you've had your gallbladder removed
- you suffer chronic stress
- you have a family history of thyroid dysfunction or disease, or autoimmune conditions
- you have been diagnosed with adrenal fatigue or another condition involving the endocrine system, or you have been previously diagnosed with an autoimmune condition
- symptoms of an underactive thyroid resonate with you yet you are told your blood test results are fine (though they tend to be skewed to one end of the 'normal' range)
- emotionally, you wonder when it will be your turn – when will you be able to do what you want to do, rather than what others want or need from you? (This is an observation of 18 years of clinical practice, and is not scientifically validated.)

Habits to better support your thyroid function:

- Trial a grain-free diet for a minimum of four weeks. A focus on gut health often needs to be at the heart of thyroid treatments, particularly if there are antibodies present in the blood.
- Have an experienced health professional assess your diet for iodine and selenium intake, and have a blood test to examine not only thyroid parameters, but also iron status, as these three minerals are essential for optimal thyroid function.
- Support liver and gallbladder function to assist with bowel elimination; globe artichokes are particularly good in this instance.
- Get your sex-hormone levels tested by an experienced health professional, as an excessive amount of oestrogen can interfere with good thyroid function.
- Adrenal support is almost always essential, especially when beginning to treat the thyroid.
- Explore taking a four-week break from caffeine and observe how you feel at the end of this period; drink green tea (which has a low caffeine level), herbal tea or dandelion tea instead.
- If you have a diagnosed thyroid condition and you are on synthetic medication but your symptoms are still present, explore transitioning to whole thyroid extract under medical supervision.

POSTNATAL DEPLETION

Before I begin, I want to be up-front and say that I am not a mother. I am writing this section as a health professional who has worked with thousands of mothers seeking help with their health and energy. The type of exhaustion I'm talking about here is not just the sleep deprivation that can occur with little ones in the house, but the physical and emotional depletion that pregnancy and lactation can create if the right supports – both nutritional and emotional – are not in place, and the concurrent journey of that type of depletion alongside the sleep deprivation and worry.

Dr Oscar Serrallach, a family practitioner based in rural Australia, describes postnatal depletion as a common phenomenon of fatigue and exhaustion combined with a feeling of ongoing 'baby brain'. Baby brain is a term that encompasses the symptoms of poor concentration, poor memory and emotional lability (when your emotions fluctuate much more easily than previously, such as if you're crying for no reason). There is often a feeling of isolation, vulnerability, and of not feeling 'good enough'. This is experienced by many mothers, and is an understandable and, at times, predictable outcome associated with the extremely demanding task of being a mother. Along with these features, Dr Serrallach has identified a typical associated biochemical 'fingerprint' that is partly the cause of, and partly the result of, postnatal depletion, and is separate from postnatal depression. Symptoms include:

- fatigue and exhaustion
- being tired on waking
- falling asleep unintentionally
- hyper-vigilance (a feeling that the 'radar' is constantly on), which is often associated with feeling anxious or a sense of unease
- feeling tired but wired
- a sense of guilt or shame around the role of being a mother
- a loss of self-esteem, which is often associated with a sense of isolation and apprehension or a fear of socialising or leaving the house
- frustration, a sense of being overwhelmed, and a sense of not coping are also common.

Some women will read this list and say 'That's how it is for *all* mothers', but that's not true. Not everyone feels that way – at least not constantly – and if you do, it is important that you seek support, both nutritional and emotional.

WHAT CAUSES POSTNATAL DEPLETION AND WHAT TO DO ABOUT IT

What creates the chemistry that allows post-natal depletion and the unrelenting fatigue that comes with it? There is no single factor at work here, so let's look at some of the main ones and what you can do to address them.

The tolls of pregnancy

Physically, the process of 'growing' a baby exacts quite a toll. The placenta passes nearly 7 grams of fat a day to the growing baby at the end of the pregnancy term – a much higher amount than in any other mammal. Furthermore, about 60 per cent of the total energy that goes to the baby via the placenta is to feed the brain. For other primates, including gorillas, the figure is closer to 20 per cent.

The placenta serves both the baby and the mother. During the pregnancy, the mother supplies everything that the growing baby needs, which is why so many mothers become low in iron, zinc, vitamin B12, folate, iodine and selenium. They also have much lower reserves in important omega-3 fats like DHA, and specific amino acids from proteins.

Studies suggest that, on average, a mother's brain shrinks approximately 5 per cent in the prenatal period, partly to help supply the baby's needs, but also as part of the physical and emotional rewiring that occurs to equip women for motherhood. It supports the creation of a 'baby radar', where mothers become intuitively aware of their child's needs, for example whether they are cold or hungry or if they cry at night.

However, this hyper-vigilance can become dangerous for the mother if she, in turn, is not supported. While there is plenty of prenatal support available, as soon as a baby is born the whole focus goes to the child. There is often very little focus on the mother, who disappears into the shadows of her new role, which is often a more difficult transition if the woman has come away from a working role where she felt highly capable. This new hyper-vigilance is obviously vital for the survival of the child, but if the mother is living in an unsupportive household or society, it can lead to sleep problems, self-doubt, insecurity, feelings of unworthiness and a concerning, dangerous level of fatigue.

Sleep deprivation

The sleep deprivation that comes with having a newborn can itself be depleting. Some research suggests that in the first year the average sleep debt is around 700 hours. Prioritising sleep is a conundrum for many mothers, as they are too tired, too stressed and too busy to sleep well. Yet they must, as it is a cornerstone of wellbeing and recovery from depletion. If you sleep quite well but your sleep is interrupted because little ones need you in the night, please remember that they are young for only a short amount of time and there will come a time when they sleep through the night. Worrying about your sleep, and disruptions that you can't do anything about, only serves to disrupt your sleep further. So I encourage you to accept that right now your sleep will be disrupted, and make the most of the nights when it is not! However, 'sleep-support' houses now exist and allow new, sleep-deprived mums to sleep there, while nurses keep an eye on the baby, at the other end of the house. Make full use of these services!

Lack of support

One of the biggest factors in the rise of postnatal depletion in our society is the shift away from family and societal support. Multigenerational support groups for mothers have been part of indigenous cultures for millennia, although they are sadly absent in our culture. Now, too many women do everything on their own.

If you feel as though you may be experiencing postnatal depletion, let a friend know and seek professional support. Psychologists specialising in supporting new mums suggest that the primary relationship between the mother and her co-parent (if she has one) often needs some special attention during the early childhood years. There are psychologists and therapists who specialise in this type of 'relationship rebuilding', and it can be beneficial to every element of the family dynamic.

The advent of the 'super mum'

In Australia, the average age for a mother having her first baby is 30.9 years, and too many women are – from a physical health perspective – going into motherhood in a depleted state, trying to blend careers, social schedules, family needs and the chronic sleep deprivation that has become the norm in our society.

There is also a perceived notion that a mother has to be everything to everyone (the super mum), something I have discussed in my book *Rushing Woman's Syndrome*. As a result, many mothers suffer in silence and are not receiving the education, information or support that they need to nurture themselves.

Humans have been on the planet for about 150,000 years and it hasn't even been 100 years since women (in significant numbers) entered the workforce. Of *course* we are capable. But our body has never had to juggle so much before. This experience for our nervous system and adrenals is relatively new, and we are still adapting. (I did a TEDx talk about this in 2014, and you can watch it on YouTube if understanding more about what we are now asking of our bodies interests you.)

Numerous child and maternal health experts believe that as a society we tend not to allow mothers to fully recover after childbirth before getting pregnant again or before they go back to work. And the juggling act they have to perform each day can be overwhelming and exhausting and further deplete their nutritional and emotional resources. It is not uncommon to see the phenomenon of a mother giving birth to two children from separate pregnancies inside 18 months, particularly if they believe their 'fertility window' is rapidly closing. Also, with assisted reproduction we are seeing higher rates of twins, which can physically exacerbate any depletion, having supplied two growing babies simultaneously with nutrition. You can do it. Of course you can do it! But additional support is required. And not enough people are aware of this or are taking actions to prevent or remedy the depletion.

Inadequate nutrition

Add to this that our food is becoming increasingly nutrient-poor, whether through the depletion of nutrients from the soil or through people choosing too many packaged and processed foods. Because we are so tired, convenience often wins out over nutrition – which is why, alternatively, slow-cooker and one-pot-wonder meals can be so practical, convenient and nutritious. Addressing micronutrient deficiencies along with a balanced intake of macronutrients and opting for a wholefood way of eating with plenty of vegetables are important places to start. The most important initial micronutrients include iron, zinc, vitamin B12, vitamin C and vitamin D. Additionally, supplemental magnesium and good-quality appropriate herbal medicine can also be highly beneficial.

A CULTURAL PERSPECTIVE

So is postnatal depletion really a new thing, or has it always been around? Really, how would we know? It has certainly become common these days and with more doctors becoming aware of it, more and more women will begin to obtain the support that they need. We understand that most of the 'primitive cultures' or first peoples of the world had very specific post-partum practices to ensure that mothers made a full recovery from childbirth. This is something that is not talked about much today. From China to India, from indigenous Australians to the First Nations peoples of America, there have been centuries of very deliberate practices in nutritional recovery, spiritual cleansing and protection, as well as elaborate social supports. For example, in traditional Chinese culture they observe the sitting month 'Zuo Yue Zi' where the mother does not leave the house for 30 days, does not receive any visitors, and has no duties apart from breastfeeding the baby. Special rebuilding, warming foods are supplied, and the mother is not allowed to get cold in that time. Ancient cultures have had the realisation that Western society unfortunately has not: for society to be well and prosper, the mothers must be fully supported and healthy – in every sense of the word.

Stress

We live in a society where many men and women are in a state of ongoing stress, and if they haven't been taught how to relax or switch off or manage their fears and concerns, then this alone (as we have seen) can have a profound impact on hormones, immune function, brain structure and gut health. Women who have chosen motherhood can also experience additional stress if they have had difficulty conceiving, suffered miscarriage or stillbirth, or experienced complications during the pregnancy or birth.

Once the baby is born, some mothers (especially those with minimal support) find the responsibility of caring for a new baby (let alone themselves) quite overwhelming, and worry about their abilities as a mother and/or partner, with attached feelings of guilt and reproach. Sometimes, breastfeeding is painful, difficult or even impossible, or the baby is unwell or never sleeps. All of these situations can create stress for new mums and this is where help from experienced health professionals is invaluable. There is always help available.

If you want to take things deeper, everything – and I mean *everything* – that happens in our lives can be our teacher if we choose to see it that way. Instead of getting caught up in negative self-judgement, simply notice that you judge yourself and bring curiosity to the experience. Why are you so harsh on yourself? Why you are reacting with rage or resentment or withdrawal or immense sadness? Use the emotion as a reminder to get back in touch with truths that are welling up inside you. I say all this with love.

Restorative practices that activate the calming, parasympathetic nervous system (see page 45) are a critical part of recovery from depletion. Restorative yoga, Stillness Through Movement and acupuncture are all beautiful and highly effective. You need to make sure that you can actually relax, as I have found that many women today cannot. So, as crazy as it sounds, if you need help to learn how to relax again, then these restorative practices will be even more important.

THE WRAP UP

As we have seen, poor energy can result from a combination of factors: biochemical, nutritional and emotional. For some people, there are very specific paths of treatment, such as for those suffering adrenal fatigue, liver congestion, thyroid dysfunction or chronic infections. For others, poor energy can be addressed by simply changing what and how they eat, how they move, how they breathe or how they sleep.

The rest of this book is dedicated to how you eat, or more precisely, how you nourish yourself. However, before we head into the kitchen, please hear this. Shift any focus you have on your weight, to a genuine focus on caring for your health and energy. Cultivate great energy! It is a wondrous resource and a marker of excellent health. To do this though, you have to believe that you are worth taking care of so live your life in touch with how precious it is and how precious you are, and treat yourself accordingly.

'You have one body,
and only you can
be responsible for it.'

PART TWO

RECIPES FOR ENERGY

CHAPTER ONE

BREAKFAST

CRUNCHY BUCKWHEAT MUESLI

Serves:
6

Preparation time:
10 minutes, plus overnight soaking

Cooking time:
50 minutes

Prep ahead:
- Soak buckwheat, almonds and seeds

1 cup (200 g) buckwheat (see Glossary)
½ cup (80 g) raw almonds, chopped
3 tablespoons sunflower seeds
3 tablespoons pumpkin seeds
1 tablespoon sesame seeds
2 tablespoons extra virgin coconut oil, melted
1 teaspoon ground cinnamon
½ teaspoon ground ginger
¼ teaspoon ground turmeric (see Glossary)
1 tablespoon pure maple syrup (optional)
milk of your choice, to serve
sliced banana and blueberries, to serve

This delicious crunchy muesli uses very little sweetener and is a nutrient-dense alternative to many packaged breakfast options on the shelves. The combination of seeds gives it a wonderful mineral boost. I find many people love it served with banana and blueberries, but you can use any fruit you like.

Place the buckwheat, almonds and seeds in a medium bowl, cover with cold water and leave to soak overnight. The next morning, rinse and drain well.

Preheat the oven to 150°C (300°F) and line a baking tray with baking paper.

Tip the soaked buckwheat mixture into a bowl, add the coconut oil, spices and syrup, if using, and mix well to coat.

Spread the buckwheat mixture over the prepared tray. Bake for 50 minutes or until lightly golden and dry. Set aside to cool completely.

Serve with your choice of milk, topped with sliced banana and blueberries.

COCONUT RICE BIRCHER

Serves:
4

Preparation time:
10 minutes, plus overnight soaking

Prep ahead:
- Cook rice and cool
- Mash banana
- Toast coconut

1 cup (160 g) cooked brown rice, cooled
1 cup (250 ml) organic coconut water (see Glossary)
1 teaspoon vanilla extract
1 tablespoon chia seeds
2 tablespoons pumpkin seeds
2 tablespoons sunflower seeds
½ teaspoon ground cardamom
½ teaspoon ground cinnamon
1 × 400 ml tin coconut cream
2 ripe bananas, mashed
2 tablespoons toasted coconut flakes

This is a nourishing version of rice pudding – the coconut cream and seeds will help to fuel you for longer. You can leave out the mashed banana if you'd prefer to have a fruit-free option, or substitute a handful of fresh or frozen berries. Leftovers will keep in an airtight container in the refrigerator for up to 5 days.

Put the cooked rice, coconut water, vanilla, chia seeds, pumpkin seeds, sunflower seeds, cardamom, cinnamon and half the coconut cream in a large bowl and mix well to combine. Cover and leave in the refrigerator overnight.

The next day, stir through the mashed banana.

Serve the coconut rice bircher in bowls, topped with the remaining coconut cream and the coconut flakes.

CREAMY BUCKWHEAT & ALMOND PORRIDGE

Serves:
4

Preparation time:
15 minutes, plus overnight soaking

Prep ahead:
- Soak buckwheat and almonds

1 cup (200 g) buckwheat (see Glossary)
1 cup (160 g) raw almonds
2 small red apples, unpeeled, coarsely grated
1 teaspoon vanilla extract
2 teaspoons finely chopped orange zest
1 cup (250 ml) freshly squeezed orange juice
1 tablespoon hulled tahini, at room temperature (see Glossary)
1 teaspoon ground cinnamon
½ teaspoon ground ginger
200 g fresh berries of your choice
1 tablespoon pumpkin seeds
1 teaspoon raw honey (optional)

This beautiful warming breakfast is perfect for crisp autumn and winter mornings. If you'd like a lower fructose option, you can swap the orange juice for water or your choice of milk. Any leftover porridge can be stored in an airtight container in the refrigerator for up to 3 days.

Place the buckwheat and almonds in a bowl, cover with cold water and leave to soak overnight. Rinse and drain well.

Transfer the buckwheat and almonds to a food processor or blender. Add the apple, vanilla, orange zest and juice, tahini and spices and process until smooth and creamy, scraping down the side of the processor or blender a few times.

Spoon the buckwheat porridge into shallow bowls and top with berries, pumpkin seeds and a drizzle of honey, if using.

SIX QUICK BREAKFASTS

Each recipe serves 4

PUMPKIN QUINOA PORRIDGE

1 cup (250 g) mashed roasted pumpkin
½ teaspoon ground cinnamon
2 tablespoons extra virgin coconut oil, melted, or ghee (see Glossary)
2 cups (370 g) rinsed, cooked quinoa (see Glossary)
2 tablespoons almond butter (see Glossary)
100 g blueberries
1 pomegranate, seeds removed
1 tablespoon pure maple syrup (optional)

Place the pumpkin, cinnamon and half the coconut oil or ghee in a saucepan and cook over medium heat for 5 minutes or until warmed through.

Place the quinoa and remaining coconut oil or ghee in a separate saucepan and warm gently over medium heat.

Spoon the quinoa into a bowl and top with the pumpkin. Add the almond butter, blueberries, pomegranate seeds and a drizzle of maple syrup if you like.

BREAKFAST BOWL

4 eggs, at room temperature
1 avocado, cut into quarters
200 g cherry tomatoes, halved
2 handfuls of watercress or baby spinach leaves
2 tablespoons freshly squeezed lemon juice
2 tablespoons extra virgin olive oil
⅓ cup (55 g) mixed seeds, such as linseeds, pumpkin seeds and sunflower seeds

Bring a saucepan of water to the boil and gently lower the eggs into the water. Cook for 6 minutes, then remove from the pan and rinse under cold water. Peel straight away.

Divide the avocado, tomato and watercress or spinach among four bowls. Top with a soft-boiled egg, drizzle with lemon juice and olive oil, and sprinkle with seeds.

STRAWBERRY CHIA POTS

250 g strawberries, hulled and halved
1 banana, chopped
½ teaspoon ground cinnamon
½ teaspoon ground ginger
1 teaspoon vanilla extract
1 tablespoon melted extra virgin coconut oil, at room temperature
1 × 400 ml tin coconut cream
½ cup (70 g) black chia seeds, plus extra for sprinkling

Place 200 g of the strawberries and the banana, spices, vanilla, coconut oil and coconut cream in a blender and blend until smooth. Transfer to a bowl and fold through the chia seeds.

Divide among four serving pots or bowls, then cover and refrigerate overnight. Serve with the remaining strawberry halves and a sprinkle of extra chia seeds.

>>

SESAME EGGS WITH ASPARAGUS SOLDIERS

1 bunch asparagus (180 g), trimmed
1 tablespoon Dukkah (see page 166)
 or toasted sesame seeds
sea salt
4 eggs, at room temperature

Steam the asparagus for 3 minutes or until bright green and tender.

Place the dukkah or sesame seeds in a bowl, add salt and mix to combine.

Bring a saucepan of water to the boil. Gently lower the eggs into the pan and cook for 4 minutes. Remove from the pan and transfer to egg cups.

Cut the tops off the eggs straight away, sprinkle with the dukkah or sesame seeds and serve with the steamed asparagus.

AVOCADO & EGG BOWLS

1 tablespoon freshly squeezed lemon juice
1 tablespoon extra virgin olive oil
4 hard-boiled eggs, peeled and chopped
1 tomato, chopped
handful of small basil leaves, torn
handful of small flat-leaf parsley leaves, torn
sea salt and freshly ground black pepper
2 avocados, halved and stones removed

Combine the lemon juice and olive oil.

Place the egg, tomato and herbs in a bowl, season with salt and pepper and gently mix to combine.

Fill the avocado halves with the egg mixture and finish with a drizzle of the dressing.

OMELETTES WITH AVOCADO & LEAFY GREENS

1 tablespoon olive oil
4 eggs, lightly beaten
2 sheets nori, sliced into 1 cm strips
2 avocados, thinly sliced
2 handfuls of baby spinach
1 tablespoon freshly squeezed lemon juice
sea salt and freshly ground black pepper

Heat half the olive oil in a non-stick frying pan over medium heat. Add half the beaten egg and swirl to coat the base of the pan; you want a thin omelette so pour off the excess if you add too much.

Cook until the edge of the omelette starts to set, lifting it with a spatula and allowing any uncooked egg to run underneath. Once the base is set, flip and cook the other side. Remove the omelette from the pan and repeat with the remaining beaten egg.

Top each omelette with nori, sliced avocado and spinach. Drizzle with lemon juice and season with salt and pepper, then roll up to enclose the filling. Cut the omelettes in half and divide among four plates to serve.

VITAL GREENS & SALMON SCRAMBLE

Serves:
4

Preparation time:
15 minutes

Cooking time:
20 minutes

Prep ahead:
- Chop spring onions and dill
- Break salmon into pieces

3 tablespoons tricolour quinoa, rinsed well (1 cup cooked; see Glossary)
2 tablespoons olive oil, plus extra to drizzle
1 bunch asparagus (180 g), thinly sliced
4 cups (320 g) shredded silverbeet
8 eggs
2 spring onions, thinly sliced
2 tablespoons chopped dill
freshly ground black pepper
200 g hot-smoked salmon, broken into pieces (see Glossary)
2 tablespoons gomasio (see Glossary) or toasted sesame seeds

I'm a big fan of serving up vegetables as part of any meal, but there's no better way to start the day than with this combination! It's equally delicious without salmon and can be eaten warm or cold. It also makes a beautifully balanced lunch, served with some additional greens.

Cook the quinoa in a saucepan of boiling water for 15 minutes or until the tails start to separate from the grain. Drain well and set aside.

Meanwhile, heat the olive oil in a non-stick frying pan over medium heat, add the asparagus and silverbeet and cook for 3–5 minutes until the asparagus is tender but still bright green.

Whisk the eggs in a large bowl, add the spring onion and dill and season to taste with pepper.

Add the egg mixture to the pan and cook, stirring gently, until the egg is just beginning to set, then use a spatula to push the set edge into the middle of the pan. Continue until all the egg is just set.

Add the hot-smoked salmon and quinoa and cook for 2 minutes to heat through. Serve topped with gomasio or toasted sesame seeds, and extra oil to drizzle.

WINTER SPICE BUCKWHEAT PORRIDGE WITH FIG COMPOTE

Serves:
4

Preparation time:
10 minutes

Cooking time:
30 minutes

Prep ahead:
- Grate ginger
- Make fig compote

½ cup (100 g) buckwheat (see Glossary)
1 teaspoon finely grated ginger
1 teaspoon ground cinnamon
½ teaspoon ground cardamom
½ teaspoon ground turmeric (optional; see Glossary)
2 cups (500 ml) water
1 teaspoon ghee (see Glossary)
2 cups (500 ml) milk of your choice
finely grated zest of 1 orange

Fig compote
8 dried figs
1 cup (250 ml) freshly squeezed orange juice
1 teaspoon vanilla extract

Despite its name, buckwheat is not related to wheat. Not only is it gluten-free, but it has more protein than rice, wheat and corn. It is also high in the essential amino acids lysine and arginine, in which cereal crops are often deficient. You can omit the fig compote if you're keeping fruit sugars low.

To make the fig compote, place all the ingredients in a saucepan. Bring to the boil, then reduce the heat and simmer for 10 minutes or until the figs are nice and plump and most of the liquid has been absorbed.

Meanwhile, place the buckwheat, ginger, spices and water in a saucepan. Bring to the boil, then reduce the heat and simmer for 15 minutes or until the buckwheat is soft.

Add the ghee and half the milk and cook until heated through.

Divide the buckwheat porridge among four bowls and top with the remaining milk and a spoonful of the fig compote. Finish with a sprinkling of orange zest and serve.

SLOW-SIMMERED BROWN RICE CONGEE WITH SEEDS

Serves:
4

Preparation time:
15 minutes

Cooking time:
8 hours or preferably overnight

Prep ahead:
- Toast rice
- Make tamari seed sprinkle

2 tablespoons shiro miso (see Glossary)
1 ½ cups (375 ml) hot water
2 tablespoons ghee (see Glossary)
3 tablespoons gomasio (see Glossary) or toasted sesame seeds
3 tablespoons Tamari seed sprinkle (see page 188)
2 sheets nori, torn
2 spring onions, sliced

Congee
1 cup (220 g) medium-grain brown rice
6 spring onions, sliced
1 tablespoon thinly sliced ginger
1 litre good-quality chicken stock
2 litres water
2 tablespoons tamari (see Glossary)

In many Asian countries it is common to have a savoury breakfast. I enjoy this, too. Seeking nutritional inspiration from other cultures also tends to increase the variety in our meals. Try this breakfast and notice if you feel fuller for longer and if your energy is more sustained.

To make the congee, toast the rice in a large frying pan over medium heat for 5 minutes or until browned. You will smell when it's ready – it will have a nice toasty fragrance.

Transfer the rice to a slow cooker, add the spring onion, ginger, stock, water and tamari. Turn the slow cooker to high and leave for a minimum of 8 hours, or preferably overnight, to cook.

The next day, blend the miso with the hot water in a small saucepan over medium heat. Cook without boiling for 5 minutes.

Spoon the congee into bowls and ladle the miso broth evenly over the top. Finish with the ghee, gomasio or sesame seeds, tamari seeds, nori and spring onion.

GREEN EGGS WITH AVOCADO

Serves:
4

Preparation time:
10 minutes

Cooking time:
10 minutes

- 2 tablespoons ghee (see Glossary) or olive oil
- 2 cups (60 g) baby spinach leaves
- 8 eggs, lightly beaten
- 2 tablespoons chopped dill, plus extra sprigs, to serve
- sea salt and freshly ground black pepper
- 2 avocados, cut into large pieces
- finely grated zest and juice of 1 lime

Eggs are a dietary mainstay for me and I'm sure many of you would relate to the saying 'If there is an egg in the house, there is a meal in the house'. Eggs are a source of complete protein, meaning they contain all the essential amino acids necessary in correct proportions to support biological function. Eggs and avocado are a match made in breakfast – or lunch, or dinner! – heaven.

Heat the ghee or olive oil in a large deep frying pan over medium heat, add the spinach and cook for 2 minutes or until just wilted.

Add the egg and sprinkle the dill over the top, then season to taste with salt and pepper. Cook, stirring gently, until the egg is just beginning to set, then use a spatula to push the set edge into the middle of the pan. Continue until all the egg is just set.

Remove the egg from the pan, fold through the avocado, lime zest and juice, then serve scattered with extra dill sprigs.

TURMERIC, SWEET POTATO & PECAN BREAKFAST MUFFINS

Makes:
10

Preparation time:
20 minutes

Cooking time:
40 minutes

2 cups (240 g) almond meal
3 tablespoons desiccated coconut
3 tablespoons chia seeds
2 teaspoons baking powder
1 teaspoon bicarbonate of soda
1 cup (120 g) grated sweet potato
6 medjool dates, pitted and chopped (see Glossary)
1 teaspoon ground turmeric (see Glossary)
1 teaspoon ground cinnamon
2 eggs, lightly beaten
2 ripe bananas, mashed (you'll need about 1 cup)
2 tablespoons olive oil
½ cup (125 ml) coconut milk or milk of your choice
30 g pecans, roughly chopped

Turmeric has long been used in cooking as well as herbal medicine. Known for its bright orange colour, it has potent anti-inflammatory effects and supports liver detoxification. Turmeric is coupled with sweet potato in these muffins to make a great nutrient-dense breakfast or snack. The muffins can be frozen in airtight containers and thawed as needed.

Preheat the oven to 200°C (400°F). Line 10 × ⅓ cup (80 ml) muffin holes with paper cases.

Place the almond meal, coconut, chia seeds, baking powder, bicarbonate of soda, sweet potato, dates and spices in a large bowl, stir to combine, then make a well in the centre.

Whisk together the eggs, banana, olive oil and milk, then pour into the dry ingredients and fold through until just combined.

Divide the mixture evenly among the paper cases and top with the chopped pecans. Bake for 30–40 minutes until risen and a skewer inserted in the centre comes out clean. Cool in the tin for a few minutes, then transfer to a wire rack to continue cooling. Serve warm or at room temperature.

FLOUR-FREE BANANA PANCAKES

Serves:
4

Preparation time:
10 minutes, plus 30 minutes resting

Cooking time:
15 minutes

Prep ahead:
- Make banana pancake batter
- Toast coconut

3 ripe bananas, mashed
6 eggs, lightly beaten
1 cup (120 g) almond meal
3 tablespoons chia seeds
2 tablespoons extra virgin coconut oil

Filling
2 bananas, sliced
2 tablespoons nut butter (see Glossary)
1 teaspoon ground cinnamon
2 tablespoons toasted shredded coconut
1 tablespoon raw honey (optional)

These are sure to become a family favourite! To make a delicious 'chocolate' version of this recipe, add 1 tablespoon of raw cacao powder to the batter. Any leftover pancakes can be frozen for another day.

Place the mashed banana, egg, almond meal and chia seeds in a food processor or blender and whiz until smooth. Transfer the batter to a jug, then cover and leave to rest for 30 minutes – the batter will thicken on standing.

Heat 1 tablespoon of the coconut oil in a large non-stick frying pan over medium heat. Working in batches, add 3 tablespoons of the batter for each pancake and cook for 2 minutes or until bubbles appear on the surface.

Turn and cook the other side for 2 minutes or until golden. Remove from the pan and keep warm. Repeat with the remaining oil and batter to make 12 pancakes in total.

Layer the pancakes with sliced banana, nut butter and a sprinkling of cinnamon and coconut. Finish with a drizzle of honey, if using.

WEEKEND PREP AHEAD #1

Preparing food for the week ahead means you can step into a new week feeling organised. Shift your thinking from 'It's just another thing I have to do' to viewing it as another form of self-care, just as you would exercise or rest.

BEANS, PULSES, QUINOA & BUCKWHEAT

Prepping these foods in large batches saves you time and means you can purchase them in bulk, which will save you money.

For beans and pulses (such as adzuki beans, black beans, butter beans, cannellini beans, chickpeas, green split peas):

Place the beans or peas into a bowl, cover with cold water and leave to soak overnight. The next morning, rinse and drain well.

Bring a saucepan of water to the boil, add the beans or peas and cook for 30 minutes or until they are just tender.

Pack the cooled beans or peas into airtight containers and store in the refrigerator for up to 5 days.

For quinoa:
Rinse under cold running water and drain well. Cook in a saucepan of boiling water for 15 minutes or until the tails start to come away from the seeds. Rinse and drain well, then cool. Pack the cooled quinoa into an airtight container and store in the refrigerator for up to 5 days.

For buckwheat:
Place in a bowl, cover with cold water and leave to soak overnight. The next morning, rinse and drain well. Bring a saucepan of water to the boil, add the buckwheat and cook for 10–15 minutes. Pack the cooled buckwheat into airtight containers and store in the refrigerator for up to 5 days.

ON-THE-FLY FRITTATAS

Makes:
12

Preparation time:
20 minutes, plus overnight soaking

Cooking time:
50 minutes

Prep ahead:
- Soak and cook beans
- Grate zucchini and shred silverbeet
- Slice spring onions

3 tablespoons dried black beans
olive oil, for greasing
10 eggs
1 zucchini, grated
3 spring onions, sliced
1 cup (40 g) chopped soft herbs (such as basil, dill, flat-leaf parsley or chives)
1 cup (80 g) shredded silverbeet
50 g cherry tomatoes, chopped
sea salt and freshly ground black pepper

These frittatas are the perfect alternative to grabbing a muffin or pastry on the way to work, and they will certainly make you feel more energised. They can be made in advance and eaten warm or cold. These frittatas also freeze well so are great to have on hand. Simply thaw and warm through, or serve at room temperature.

Place the beans in a bowl, cover with cold water and leave to soak overnight. The next morning, rinse and drain well.

Bring a saucepan of water to the boil, add the beans and cook over high heat for 30 minutes or until the beans are just tender – do not overcook or they will turn to mush when you mix them through the frittata.

Preheat the oven to 180°C (350°F). Grease 12 x ⅓ cup (80 ml) non-stick muffin holes generously with olive oil.

Place the eggs in a bowl and whisk to combine. Add the zucchini, spring onion, herbs, silverbeet, tomatoes and beans and season with salt and pepper. Mix until just combined.

Divide the mixture evenly among the muffin holes. Bake for 20 minutes or until the frittatas are set and start to come away from the sides of the muffin holes. Serve warm or at room temperature.

MISO BROWN RICE, SPINACH & EGG SWIRL SOUP

Serves:
4

Preparation time:
10 minutes

Cooking time:
15 minutes

Prep ahead:
- Slice spring onions
- Break mushrooms
- Grate ginger

4 spring onions, thinly sliced
6 dried shiitake mushrooms, broken into pieces (see Glossary)
½ cup (80 g) cooked brown rice
1 tablespoon grated ginger
1 litre boiling water
2 tablespoons shiro miso (see Glossary)
1 cup (30 g) baby spinach leaves, shredded
2 eggs, lightly beaten
1 tablespoon tamari (see Glossary)

This Asian-inspired breakfast is a great alternative if you're a bit bored with your current breakfast options. It is particularly nourishing in the cooler months, especially if you feel the cold. It's also a good way to start the day if you don't feel overly hungry in the morning.

Place the spring onion, mushroom, rice, ginger and water in a large saucepan and bring to the boil, then reduce the heat and simmer for 10 minutes.

Blend the miso with 2 tablespoons of the hot liquid, then add to the pan but do not allow it to boil.

Stir in the spinach and egg and cook, stirring, for 3 minutes or until the egg sets. Add the tamari and mix to combine, then serve.

CHAPTER TWO

LUNCH

MEXICAN CORN FRITTERS WITH TOMATO & AVOCADO SALSA

Serves:
4

Preparation time:
15 minutes

Cooking time:
15 minutes

Prep ahead:
- Remove kernels from corn cobs

2 corn cobs, kernels removed (2 cups/400 g)
1 teaspoon sweet smoked paprika
1 teaspoon ground coriander
½ teaspoon dried thyme
finely grated zest of 1 lime
3 large eggs, lightly beaten
3 tablespoons olive oil

Tomato & avocado salsa
1 large ripe avocado, roughly smashed
½ small red onion, finely chopped
1 tablespoon freshly squeezed lime juice
2 ripe tomatoes, finely chopped
2 tablespoons chopped coriander leaves
sea salt

Corn comes in different colours with varying nutritional benefits. In the case of yellow corn, it's the density of the (antioxidant) carotenoids that stand out. Fritters are a fabulous way to incorporate more vegies in the diet, particularly with fussy eaters. The delicious tomato and avocado salsa can be served with nearly everything!

To make the salsa, place the avocado, onion and lime juice in a bowl and roughly mash to combine. Add the remaining ingredients and mix well. Set aside.

Place the corn, paprika, ground coriander, thyme and lime zest in a bowl. Add the egg and whisk until well combined.

Heat half the olive oil in a large deep non-stick frying pan over medium heat. Add heaped tablespoons of the batter in batches and cook for 3 minutes on each side or until cooked through. Remove and cover to keep warm while you cook the remaining mixture, adding the remaining oil as needed.

Serve the fritters warm topped with the salsa.

NICOISE SALAD

Serves:
4

Preparation time:
25 minutes, plus overnight soaking

Cooking time:
30 minutes

Prep ahead:
- Soak and cook butter beans
- Cook potatoes
- Hard-boil eggs
- Make dressing

½ cup (100 g) dried butter beans (see Glossary)
100 g baby chat potatoes
250 g green beans, trimmed
250 g cherry tomatoes, halved
100 g pitted kalamata olives, sliced
6 red radishes, thinly sliced
handful of basil leaves, torn
4 hard-boiled eggs, peeled

Dressing
⅓ cup (60 g) capers, rinsed and roughly chopped
1 tablespoon dijon mustard (preferably organic)
1 garlic clove, crushed
2 tablespoons freshly squeezed lemon juice
2 teaspoons raw honey
⅓ cup (80 ml) extra virgin olive oil
sea salt and freshly ground black pepper

I love potato and egg together. Combine that with a beautiful lemon and mustard dressing and you have a real winner. This is a great recipe to make in advance and take to work.

Place the butter beans in a bowl, cover with cold water and leave to soak overnight. The next morning, rinse and drain well. Cook the beans in a saucepan of boiling water for 30 minutes or until they are just soft, then rinse and drain well.

Meanwhile, steam the potatoes for 10 minutes or until tender. Allow to cool slightly, then cut into thick slices. Steam the beans for 3–5 minutes until tender and bright green.

Place the butter beans, potato slices, green beans, tomato, olives, radish and basil in a bowl and gently mix to combine.

To make the dressing, whisk together all the ingredients. Drizzle half the dressing over the vegetables and gently toss to coat.

Pile the salad onto four plates and grate the hard-boiled eggs over the top. Drizzle with the remaining dressing and serve.

RED QUINOA, SEED & PECAN SALAD WITH LEAFY GREENS

Serves:
4

Preparation time:
15 minutes, plus cooling time

Cooking time:
15 minutes

Prep ahead:
- Cook quinoa
- Roast pecans
- Make dressing

½ cup (100 g) red quinoa (see Glossary), rinsed well
100 g salad leaves
50 g snow pea sprouts
1 cup (120 g) frozen peas, thawed
1 cup (20 g) mixed soft herb leaves, such as dill, mint and basil
2 zucchini, peeled into ribbons
80 g roasted pecans
3 tablespoons toasted pumpkin seeds, sunflower seeds and linseeds

Dressing
2 tablespoons dijon mustard (preferably organic)
1 garlic clove, crushed
1 tablespoon balsamic vinegar
1 teaspoon pure maple syrup
pinch of dried chilli flakes (optional)
3 tablespoons extra virgin olive oil

Quinoa is one of the few plant foods that can be considered a complete protein, providing all of the essential amino acids. It is also a helpful and nourishing addition to a gluten-free diet. You could serve this salad with an egg or your choice of protein for a more substantial option.

Cook the quinoa in a saucepan of boiling water for 15 minutes or until the tails start to come away from the seeds. Rinse and drain well, then set aside to cool.

Transfer the cooled quinoa to a bowl, add the salad leaves, sprouts, peas, herbs, zucchini and pecans and mix to combine.

To make the dressing, whisk together all the ingredients. Pour over the quinoa salad and toss to coat.

Pile the salad onto a plate, sprinkle the seeds over the top and serve.

SPICED CHICKEN & LENTIL SALAD WITH TAHINI CREAM & SMASHED AVOCADO

Serves:
4

Preparation time:
20 minutes, plus
10 minutes resting

Cooking time:
15 minutes

Prep ahead:
- Marinate chicken
- Cook lentils
- Make tahini cream

2 organic chicken breast fillets (600 g)
1 tablespoon harissa spice mix
3 tablespoons olive oil
½ cup (110 g) black beluga or puy-style lentils (see Glossary)
sea salt and freshly ground black pepper
2 ripe avocados
1 tablespoon freshly squeezed lemon juice
100 g baby rocket

Tahini cream
½ cup (140 g) hulled tahini, at room temperature (see Glossary)
½ teaspoon ground cumin
3 tablespoons freshly squeezed orange juice
1 tablespoon freshly squeezed lemon juice
large handful of coriander leaves
½ cup (125 ml) water
sea salt and freshly ground black pepper

Here, beautiful flavours combine to offer you a nutritionally valuable lunch. Tahini and avocado are both sources of nourishing fats that will help to fuel you into the afternoon. Serve with another big handful of leafy greens such as rocket and spinach to further boost the goodness.

Preheat a barbecue flat plate, if using.

Place the chicken breasts in a bowl, add the harissa spice mix and half the olive oil and toss to coat well in the spice mixture. Cook the chicken on the preheated flat plate or in a frying pan over medium–high heat for 15 minutes, turning once. Insert a skewer into the thickest part of the breast – if the juices run clear, the chicken is cooked. Remove, then cover and rest for 10 minutes before cutting into thick slices.

Meanwhile, cook the lentils in a saucepan of boiling water for 15 minutes or until just soft – do not overcook or they will be mushy. Drain well, then stir in the remaining olive oil and season well with salt and pepper.

Roughly mash the avocados with the lemon juice.

To make the tahini cream, place the tahini, cumin, orange juice, lemon juice and coriander in a food processor and process to combine. Add the water gradually until thinned to your preferred consistency. Season to taste with salt and pepper.

To serve, spread a spoonful of the tahini cream onto each plate and top with the lentils, rocket, smashed avocado and chicken.

SHREDDED CABBAGE, PEAR & HAZELNUT SALAD

Serves:
4

Preparation time:
30 minutes, plus overnight soaking

Prep ahead:
- Make cashew mayonnaise
- Shred cabbage
- Make dressing
- Roast hazelnuts

½ small savoy cabbage, finely shredded
1 small radicchio, leaves separated and torn into small pieces
4 celery stalks, cut diagonally into long thin strips
2 pears, cored and thinly sliced
½ cup (60 g) roasted hazelnuts, skins removed, roughly chopped
2 tablespoons snipped chives
1 tablespoon chopped flat-leaf parsley
1 tablespoon chopped dill

Cashew mayonnaise
¾ cup (120 g) raw cashews
1 garlic clove, roughly chopped
2 teaspoons dijon mustard (preferably organic)
½ teaspoon freshly ground black pepper
sea salt
1 tablespoon freshly squeezed lemon juice
⅓ cup (80 ml) water

Dressing
1 tablespoon freshly squeezed lemon juice
½ teaspoon freshly ground black pepper
2 tablespoons balsamic vinegar
3 tablespoons extra virgin olive oil

This is a beautiful crisp salad. The cashew mayo is a must-have – store any leftovers in an airtight container in the refrigerator for up to 5 days. (This mayo also goes wonderfully with the fish cakes on page 148.) To roast the hazelnuts, spread out on a baking tray lined with baking paper and place in an 180°C oven for 5–10 minutes, stirring to ensure they cook evenly. Rub them in a tea towel whilst still hot to remove the skins.

To make the cashew mayonnaise, place the cashews in a bowl, cover with cold water and leave to soak overnight. The next morning, rinse and drain well. Place the cashews and garlic in a blender or small food processor and blend until smooth. Add the remaining ingredients and blend until thick and creamy.

Place the cabbage, radicchio, celery, pear, hazelnuts and herbs in a large bowl.

For the dressing, whisk together all the ingredients.

Pour the dressing over the salad and gently toss to combine. Finish with a spoonful of the cashew mayonnaise.

WEEKEND PREP AHEAD #2

SALAD DRESSINGS

Make large batches of dressing on the weekend to use all week. For each dressing, place the ingredients in a jar, seal with a lid and shake well to combine. These will keep in the refrigerator for up to 1 week.

Everyday salad dressing

½ garlic clove, crushed
2 teaspoons dijon mustard (preferably organic)
2 tablespoons white or red wine vinegar
½ cup (125 ml) extra virgin olive oil
sea salt and freshly ground black pepper

Creamy tahini dressing

½ cup (140 g) hulled tahini, at room temperature (see Glossary)
2 teaspoons dijon mustard (preferably organic)
1 tablespoon balsamic vinegar
1 garlic clove, chopped
½ cup (125 ml) water
sea salt and freshly ground black pepper

Lemon and caper dressing

⅓ cup (65 g) capers, rinsed and roughly chopped
1 tablespoon dijon mustard (preferably organic)
1 garlic clove, crushed
2 tablespoons freshly squeezed lemon juice
2 teaspoons raw honey
⅓ cup (80 ml) extra virgin olive oil
sea salt and freshly ground black pepper

Maple and chilli dressing

2 tablespoons dijon mustard
(preferably organic)
1 garlic clove, crushed
1 tablespoon balsamic vinegar
1 teaspoon pure maple syrup
pinch of dried chilli flakes (optional)
3 tablespoons extra virgin olive oil

Orange and balsamic dressing

2 tablespoons balsamic vinegar
2 tablespoons freshly squeezed orange juice
1 teaspoon dijon mustard (preferably organic)
½ cup (125 ml) extra virgin olive oil
sea salt and freshly ground black pepper

Pink dressing

2 teaspoons sumac (see Glossary)
2 tablespoons apple cider vinegar
⅓ cup (80 ml) extra virgin olive oil
sea salt and freshly ground black pepper

MAPLE-ROASTED CARROT, LENTIL & ARTICHOKE SALAD

Serves:
4

Preparation time:
20 minutes

Cooking time:
40 minutes

Prep ahead:
- Roast vegetables
- Cook lentils
- Make dressing

4 carrots, cut into thick batons
2 red capsicums, seeded and cut into strips
200 g drained bottled artichoke hearts
2 red onions, cut into wedges
1 tablespoon olive oil
1 tablespoon balsamic vinegar
1 tablespoon pure maple syrup
sea salt and freshly ground black pepper
½ cup (110 g) puy-style or green lentils (see Glossary)
100 g kale leaves, torn
3 tablespoons pumpkin seeds

Dressing
2 tablespoons apple cider vinegar
2 teaspoons dijon mustard (preferably organic)
⅓ cup (80 ml) extra virgin olive oil
sea salt and freshly ground black pepper

Artichokes are wonderful for gut health, which is vital to your ability to feel energised. They contain inulin, which is known as a prebiotic. Unlike other carbohydrates, this isn't broken down in the first segments of the digestive tract, instead passing through to the large intestine where it ferments, helping to support healthy gut flora.

Preheat the oven to 200°C (400°F) and line a large baking tray with baking paper.

Place the carrot, capsicum, artichokes and onion in a large bowl. Whisk together the olive oil, balsamic vinegar and maple syrup, then pour over the vegetables. Season with salt and pepper and toss to combine. Transfer the vegetables to the prepared tray and bake for 40 minutes or until the vegetables are soft.

Meanwhile, cook the lentils in a saucepan of boiling water for 15 minutes or until just soft. Rinse and drain well.

Combine the vegetables, lentils, kale and pumpkin seeds in a large bowl.

To make the dressing, whisk together all the ingredients. Drizzle the dressing over the salad and serve.

MARINATED MUSHROOM & BUCKWHEAT SALAD

Serves:
4

Preparation time:
20 minutes, plus
1 hour marinating

Cooking time:
10 minutes

Prep ahead:
- Marinate mushrooms
- Cook buckwheat
- Shred cabbage

100 g swiss brown mushrooms, halved
100 g button mushrooms, sliced
2 spring onions, sliced
2 tablespoons extra virgin olive oil
1½ tablespoons apple cider vinegar
1 tablespoon freshly squeezed lemon juice
2 teaspoons tamari (see Glossary)
2 teaspoons pure maple syrup
2 teaspoons dijon mustard (preferably organic)
½ cup (100 g) buckwheat (see Glossary)
¼ Chinese cabbage (wombok), finely shredded
3 tablespoons toasted mixed seeds (such as pumpkin, sesame, sunflower and linseeds)
1 tablespoon small coriander leaves
1 tablespoon small mint leaves
1 tablespoon snipped chives

Adding herbs to your meals is a wonderful way to give them a nutrient-dense flavour boost. This salad is gorgeous by itself and fulfilling in part due to the fibre content; however you can also pair it with your choice of protein if you like.

Place the mushrooms and spring onion in a bowl. In a smaller bowl, whisk together the olive oil, vinegar, lemon juice, tamari, maple syrup and mustard and pour over the mushrooms. Set aside to marinate for 1 hour.

Place the buckwheat in a saucepan of boiling water and cook for 10 minutes or until just soft – do not overcook or it will be mushy. Rinse and drain well and leave to cool.

Add the buckwheat, cabbage, seeds and herbs to the mushroom mixture and gently toss to combine. Serve at room temperature.

IN-THE-PINK SALAD

Serves:
4

Preparation time:
20 minutes

Prep ahead:
- Shred cabbage
- Grate beetroot
- Chop watermelon
- Hard-boil eggs
- Make dressing

¼ small red cabbage, finely shredded
2 beetroot, peeled and grated
2 cups (400 g) chopped seeded watermelon
handful of mint leaves
2 hard-boiled eggs, peeled and halved
3 tablespoons pumpkin seeds
3 tablespoons pomegranate seeds (optional)
micro herbs (optional), to serve

Dressing
2 teaspoons ground sumac (see Glossary)
2 tablespoons apple cider vinegar
⅓ cup (80 ml) extra virgin olive oil
sea salt and freshly ground black pepper

This gorgeous salad is perfect for summer evenings! When it comes to nutrients, the more colour you can incorporate into your meals the better. For example, here the beautiful bright colour of red cabbage comes from anthocyanins, which act as antioxidants. Couple this with leafy greens and your bowl is swimming with nutrients ready for you to utilise.

Place the cabbage and beetroot in a bowl and mix to combine.

To make the dressing, whisk together all the ingredients. Pour half over the cabbage and beetroot and toss to coat in the dressing.

Arrange the cabbage salad on four serving plates and top with the watermelon, mint, egg halves, pumpkin seeds and pomegranate seeds, if using. Sprinkle with micro herbs, if you like.

Drizzle over the remaining dressing and serve.

CHICKEN DHAL, CORN & COCONUT SOUP

Serves:
4

Preparation time:
15 minutes

Cooking time:
35 minutes

Prep ahead:
- Remove kernels from corn cobs
- Chop lemongrass
- Make coriander coconut cream mix

1 cup (210 g) split mung dhal or red lentils (see Glossary)
4 organic chicken thigh fillets (400 g), skin removed, cut into thin slices
2 corn cobs, kernels removed (2 cups/400 g)
4 spring onions, chopped
1 tablespoon grated ginger
8 kaffir lime leaves, torn (see Glossary)
1 tablespoon finely chopped lemongrass stalk, white part only
1.5 litres good-quality chicken stock
1 tablespoon extra virgin coconut oil
100 g baby spinach leaves, shredded
2 tablespoons freshly squeezed lime juice
½ cup (125 ml) coconut cream
2 tablespoons chopped coriander leaves
1 large green chilli, halved, seeded and finely chopped (optional)
1 tablespoon finely grated lime zest

This cross between a dhal and chicken-and-corn soup is fresh, creamy and quite simply delicious. Legumes are an excellent source of fibre and are high in protein, while the coconut cream is rich in dietary fat – a combination that will keep you feeling sustained for hours. This soup can be frozen in an airtight container. Just make sure you leave a little room at the top for the soup to expand as it freezes.

Place the dhal or lentils, chicken, corn, spring onion, ginger, kaffir lime leaves, lemongrass, stock and coconut oil in a large saucepan. Bring to the boil over high heat, then reduce the heat to medium and simmer for 30 minutes or until the dhal is soft and creamy.

Remove from the heat and stir in the spinach and lime juice. Cover and set aside for 5 minutes.

Meanwhile, place the coconut cream, coriander and chilli, if using, in a bowl and mix to combine.

Ladle the soup into four bowls, top with a spoonful of coriander coconut cream and finish with a sprinkling of lime zest.

CHOPPED SALAD WITH CHICKEN & AVOCADO

Serves:
4

Preparation time:
20 minutes, plus overnight soaking

Cooking time:
15 minutes

Prep ahead:
- Soak almonds
- Poach and shred chicken

½ cup (80 g) raw almonds
2 small organic chicken breast fillets (400 g)
3 spring onions
1 Lebanese cucumber
100 g cherry tomatoes
1 baby cos lettuce
100 g alfalfa sprouts
handful of snipped chives
handful of mint leaves
1 ripe avocado, diced

Dressing
1 tablespoon wholegrain mustard
1 tablespoon apple cider vinegar
2 tablespoons extra virgin olive oil
sea salt and freshly ground black pepper

This is one of those recipes you will keep coming back to! The tangy dressing perfectly complements the combination of herbs and vegetables. The addition of almonds brings a great source of B vitamins, magnesium, selenium, zinc, potassium, calcium and manganese – all vital nutrients for energy. If you'd like a vegan option, simply omit the chicken; for a vegetarian option add an egg instead.

Place the almonds in a bowl, cover with cold water and leave to soak overnight. The next morning, rinse and drain well, then slice.

Place the chicken in a deep frying pan and cover with water. Bring to the boil, then reduce the heat and simmer, covered, for 15 minutes or until the chicken is cooked and tender. Remove from the pan and allow to cool, then finely shred the meat.

Thinly slice the spring onions on a very large chopping board, add the cucumber and chop together with the spring onion. Add the tomatoes and chop. Shred the lettuce and fold into the chopped salad ingredients.

Add the alfalfa, almonds, chives and mint and chop until all the ingredients are combined. Add the chicken and diced avocado and mix well.

Make a well in the middle of the salad. Pour the dressing ingredients into the well and mix to combine, then fold the dressing into the chopped salad. You can take the salad to the table on the board, or transfer it to individual plates and serve.

SPLIT PEA SOUP

Serves:
4

Preparation time:
20 minutes, plus overnight soaking

Cooking time:
1 hour 10 minutes

Prep ahead:
- Soak peas
- Chop vegetables

1⅓ cups (310 g) green split peas
2 tablespoons olive oil
1 brown onion, chopped
3 celery stalks, sliced
2 carrots, chopped
1.5 litres water
1 bay leaf
1 cinnamon stick
1 cup (120 g) frozen peas, thawed
sea salt and freshly ground black pepper
2 tablespoons baby capers, rinsed
2 tablespoons roughly chopped dill
finely grated zest and juice of 1 lemon

This simple soup is a flavour-filled mouth festival! Cinnamon is a warming spice so this is a great option for the cooler months. It's also quite light so is a perfect meal if you're not feeling too hungry but want the nourishment. This soup can be frozen in an airtight container (make sure you leave enough room at the top for it to expand as it freezes).

Place the split peas in a bowl, cover with cold water and leave to soak overnight. The next morning, rinse and drain well.

Heat half the olive oil in a large saucepan over medium heat, add the onion and cook for 10 minutes or until soft and golden.

Add the celery and carrot and cook for 5 minutes or until they start to soften. Add the split peas, water, bay leaf and cinnamon stick. Bring to the boil, then reduce the heat and simmer for 45 minutes or until the split peas are very soft and creamy.

Add the thawed peas and cook for 5 minutes or until heated through. Set aside to cool slightly, then blend with a stick blender until smooth. Season with salt and pepper.

Heat the remaining oil in a small frying pan over medium heat, add the capers and cook for 5 minutes or until they are slightly crisp.

Ladle the soup into four bowls and serve topped with the fried capers, chopped dill, lemon zest and juice.

LEEK, ASPARAGUS & SPINACH FRITTATA

Serves:
4

Preparation time:
15 minutes

Cooking time:
45 minutes

2 tablespoons olive oil, plus extra for greasing
1 leek, white part only, thinly sliced
1 bunch asparagus (180 g), cut into 5 cm lengths
1 zucchini, halved lengthways and thinly sliced
1 cup (120 g) frozen peas, thawed
50 g baby spinach leaves
8 eggs, lightly beaten
½ cup (125 ml) milk of your choice
1 tablespoon dijon mustard (preferably organic)
3 tablespoons mixed chopped dill, mint and chives, plus extra to serve
sea salt and freshly ground black pepper

This frittata, bursting with green goodness, is a favourite of mine. I make it for breakfast, lunch or dinner. It travels well and is a great option to pack as a work lunch. It can also be frozen; simply thaw and warm through, or serve at room temperature. To boost the nutritional value even more, serve with a plant-based pesto.

Preheat the oven to 180°C (350°F). Grease the base and side of a deep non-stick 26 cm ovenproof frying pan generously with olive oil.

Heat the oil in the greased pan over medium heat, add the leek and cook for 10 minutes or until the leek is soft and golden.

Add the asparagus, zucchini and peas and cook for 5 minutes or until the zucchini has softened and there is no moisture left in the pan.

Add the spinach and cook, covered, for 3 minutes or until wilted.

Whisk together the eggs, milk, mustard and herbs, season with salt and pepper and pour the mixture into the pan. Stir once to evenly distribute the egg among the vegetables, then cook over low–medium heat for 5–10 minutes or until the base and edge are set. Transfer the pan to the oven and cook for 15 minutes or until the egg is completely set.

Slide the frittata onto a plate or board, cut into wedges and serve sprinkled with extra herbs.

SUSHI BOWL

Serves:
4

Preparation time:
20 minutes, plus overnight soaking

Cooking time:
30 minutes

Prep ahead:
- Soak cashews
- Make ginger cream
- Cook rice
- Make omelette

1 cup (220 g) medium-grain brown rice
3 cups (750 ml) water
1 tablespoon tamari (see Glossary)
3 eggs, lightly beaten
1 tablespoon toasted sesame seeds
1 tablespoon extra virgin coconut oil
1 avocado, thinly sliced
1 Lebanese cucumber, thinly sliced
1 carrot, cut into matchsticks
1 cup (120 g) frozen edamame beans, thawed
2 sheets nori
½ cup (75 g) red cabbage sauerkraut or kimchi
½ teaspoon sesame oil

Ginger cream
⅔ cup (100 g) raw cashews
¼ teaspoon fresh horseradish (see Glossary)
1 tablespoon finely grated ginger
2 teaspoons freshly squeezed lemon juice
½ cup (125 ml) water

I'm a big fan of deconstructed bowls, mainly because they save time! Here you get all the flavours and nutritional benefits of sushi without the fuss of rolling it. The ginger cream is delicious – it's a great addition to other meals, too.

To make the ginger cream, place the cashews in a bowl, cover with cold water and leave to soak overnight. The next morning, rinse and drain well. Place the cashews into a small food processor with the remaining ingredients and process until smooth and creamy.

Place the rice and water in a medium saucepan. Bring to the boil over high heat and boil for 20 minutes or until tunnels form in the rice and most of the liquid has been absorbed. Reduce the heat to low, cover and cook gently for 5 minutes or until the rice is soft and there is no liquid left in the pan. Remove from the heat, stir in the tamari and set aside to rest while you prepare the toppings.

Whisk together the egg and sesame seeds. Melt the coconut oil in a 26 cm non-stick frying pan over medium heat, add the egg mixture and cook, stirring gently, until the egg is just beginning to set, then use a spatula to push the set edge into the middle of the pan. Continue until all the egg is just set. Flip over and cook until set (this should take about 3 minutes all up), then remove from the pan and slice.

Divide the rice among four serving bowls and arrange the avocado, cucumber, carrot, edamame, nori and sesame omelette on top. Add the sauerkraut or kimchi, then drizzle over the sesame oil. Serve with the ginger cream.

RAINBOW RICE PATTIES

Makes:
12

Preparation time:
20 minutes, plus
30 minutes chilling

Cooking time:
40 minutes

Prep ahead:
- Cook rice
- Cook vegetables
- Shape patties

1 cup (220 g) medium-grain brown rice
2 tablespoons black chia seeds
3 cups (750 ml) water
1 tablespoon extra virgin coconut oil
1 cup (240 g) grated sweet potato
1 cup (240 g) grated beetroot
1 cup (50 g) finely shredded kale
2 tablespoons pumpkin seeds
2 sheets nori, finely shredded
2 tablespoons hulled tahini, at room temperature (see Glossary)
1 tablespoon shiro miso or tamari (see Glossary)
⅓ cup (50 g) toasted black or white sesame seeds
Tahini cream (see page 121), to serve (optional)
mixed soft herbs (dill, oregano, coriander leaves, flat-leaf parsley), to serve

These colourful creations also make an easy evening meal, served with protein and additional greens. The chia seeds add a source of wholefood fats including the anti-inflammatory omega-3s. Research suggests it's incredibly valuable for children to be included in vegetable preparation and this recipe is a great one to get the kids involved with. These patties are delicious on their own or served with the Tahini cream on page 121.

Place the rice, chia seeds and water in a medium saucepan. Bring to the boil over high heat and boil for 20 minutes or until tunnels form in the rice and most of the liquid has been absorbed. Reduce the heat to low, cover and cook gently for 5 minutes or until the rice is soft and there is no liquid left in the pan.

Meanwhile, melt the coconut oil in a large deep frying pan over medium heat, add the sweet potato and beetroot and cook for 5 minutes or until soft. Add the kale and cook until wilted. Transfer the vegetables to a bowl, add the chia rice, pumpkin seeds, nori, tahini and miso or tamari and use your hands to combine the mixture until it sticks together.

Line a baking tray with baking paper. Divide the mixture into 12 portions and form into patties, then press into the sesame seeds to coat on both sides. Chill for 30 minutes or as long as time permits.

Preheat the oven to 200°C (400°F). Bake the patties for 20 minutes or until crisp, golden and heated through. Serve with tahini cream, if you like, and mixed herbs.

ADZUKI BEAN, GINGER & PUMPKIN STEW

Serves:
4

Preparation time:
10 minutes, plus overnight soaking

Cooking time:
1 hour 15 minutes

Prep ahead:
- Soak beans
- Roast pumpkin

1 cup (220 g) dried adzuki beans (see Glossary)
¼ small kent pumpkin, seeded, skin left on
3 cups (750 ml) water
1 cinnamon stick
2 star anise
1 tablespoon finely grated ginger
2 spring onions, chopped
2 tablespoons tamari (see Glossary)
sea salt
handful of coriander leaves
cooked brown rice, to serve

Adzuki beans are what I consider a very restorative food. This stew is perfect for those days where you just feel a little worn down. Stews and soups are good options when you're feeling depleted, as they're gentle on the digestive system and provide nutrients that are easier for our body to utilise. The stew can be frozen for up to 1 month in an airtight container.

Place the adzuki beans in a bowl, cover with cold water and leave to soak overnight. The next morning, rinse and drain well.

Preheat the oven to 200°C (400°F).

Put the pumpkin in a roasting tin and roast for 1 hour or until very soft and golden.

Meanwhile, place the adzuki beans and water in a very large saucepan, add the cinnamon stick, star anise and ginger and cook for 1 hour or until the beans are just tender and most of the liquid has evaporated.

Add the roasted pumpkin (with the skin on) to the pan, along with the spring onion and tamari, and simmer for 15 minutes or until thickened and the pumpkin is very tender. Remove the cinnamon stick and star anise, and season to taste with salt. Scatter with coriander leaves and serve with cooked brown rice.

SUMMER DHAL

Serves:
4

Preparation time:
15 minutes, plus standing time

Cooking time:
25 minutes

Prep ahead:
- Chop broccoli
- Grate turmeric, ginger and zucchini

1 cup (210 g) split mung dhal or red lentils (see Glossary)
2 teaspoons extra virgin coconut oil
1 tablespoon grated fresh red turmeric or 1 teaspoon ground turmeric (see Glossary)
1 tablespoon grated ginger
1 teaspoon fennel seeds
1 cinnamon stick
1 brown onion, chopped
1 litre water
300 g broccoli, chopped into small florets
2 zucchini, grated
50 g baby spinach leaves
1 tablespoon freshly squeezed lemon juice
1 tablespoon ghee (see Glossary)
sea salt
2 tablespoons coriander leaves

You don't have to wait until winter to enjoy this lovely light dhal. I always encourage my patients with digestive complaints to consider incorporating more slow-cooked foods like this. It can be frozen in an airtight container for up to 1 month.

Place the dhal or lentils, coconut oil, turmeric, ginger, fennel seeds, cinnamon stick, onion and water in a pan. Bring to the boil, then reduce the heat and simmer for 15 minutes.

Add the broccoli and zucchini and cook for 10 minutes or until the dhal is very soft and creamy and most of the water has been absorbed.

Remove from the heat and discard the cinnamon stick. Stir in the spinach, lemon juice and ghee, then cover and stand for 5 minutes or until the spinach has wilted. Season to taste with salt. Serve topped with coriander.

SALMON & QUINOA FISH CAKES

Serves:
4

Preparation time:
20 minutes

Cooking time:
45 minutes

Prep ahead:
- Steam sweet potato and salmon
- Cook quinoa

300 g sweet potato, peeled and chopped
500 g skinless salmon fillet
½ cup (100 g) tricolour quinoa, rinsed well (see Glossary)
1 cup (120 g) almond meal
1 egg, lightly beaten
1 tablespoon red curry paste
100 g green beans, trimmed and thinly sliced
4 kaffir lime leaves, finely shredded (see Glossary)
finely grated zest and juice of 1 lime
2 tablespoons extra virgin coconut oil
salad greens and lime wedges (optional), to serve

Fish cakes are a great healthy staple. These have extra protein from the quinoa, meaning they're particularly satiating. Serve them with additional greens or one of the salad recipes in this book. The fish cakes can be frozen (cooked or uncooked) in an airtight container for up to 3 weeks.

Steam the sweet potato for 20 minutes or until soft. Remove from the heat and mash.

Steam the salmon for 10–15 minutes until it flakes easily. Remove from the heat and flake the salmon, discarding any bones.

Meanwhile, cook half the quinoa in a saucepan of boiling water for 20 minutes or until soft. Rinse and drain well.

Place the sweet potato, salmon, cooked quinoa, almond meal, egg, curry paste, beans, kaffir lime leaves and lime zest and juice in a bowl and mix well to combine.

Divide the mixture into four portions and form into patties. Sprinkle both sides of the patties with the remaining quinoa to lightly coat.

Melt the coconut oil in a large deep frying pan over medium heat and cook the fish cakes (in batches if necessary) for 5 minutes on each side or until golden and cooked through. Serve the fish cakes on salad leaves with lime wedges, if you like.

BRAISED CANNELLINI BEANS & GREENS FEST

Serves:
4

Preparation time:
20 minutes, plus overnight soaking

Cooking time:
1 hour 15 minutes

Prep ahead:
- Soak beans
- Chop vegetables

1 cup (220 g) dried cannellini beans
2 tablespoons olive oil
1 leek, white part only, sliced
2 garlic cloves, crushed
1 tablespoon finely chopped preserved lemon rind or finely grated lemon zest
100 g Sicilian olives, pitted
1 teaspoon ground ginger
½ teaspoon freshly ground black pepper
2 cups (500 ml) good-quality chicken stock
1 litre water
2 zucchini, sliced into rounds
1 bunch asparagus (180 g), trimmed and sliced into thirds
300 g broccoli, cut into large florets
1 cup (120 g) frozen peas, thawed
100 g sugar snap peas
2 cups (160 g) shredded silverbeet
sea salt

Dressing
1 tablespoon finely grated lemon zest
2 tablespoons extra virgin olive oil
2 tablespoons freshly squeezed lemon juice
3 tablespoons small flat-leaf parsley leaves

The lemon parsley dressing that complements this greens-packed dish makes it incredibly refreshing. Beans are a good source of dietary fibre, essential for the optimal functioning of the digestive tract, efficient elimination and helping you feel satiated.

Place the cannellini beans in a bowl, cover with cold water and leave to soak overnight. The next morning, rinse and drain well.

Heat the olive oil in a large heavy-based flameproof casserole dish over medium heat, add the leek and cook for 10 minutes or until soft and golden.

Add the garlic, lemon, olives, ginger and pepper and cook for 2 minutes or until the spices are fragrant. Add the cannellini beans, stock and water and bring to the boil, then reduce the heat and simmer, covered, for 50 minutes or until the beans are just soft.

Add the zucchini, asparagus, broccoli, peas and sugar snaps, then cover and cook for a further 10 minutes or until the vegetables are soft.

Add the silverbeet and cook for 3 minutes or until the silverbeet has wilted. Season to taste with salt.

To make the dressing, whisk together all the ingredients in a small bowl. Spoon over the braised vegetables and serve.

COCONUT STEAK SALAD WITH ROASTED SWEET POTATO, CORIANDER PESTO & LENTILS

Serves:
4

Preparation time:
20 minutes, plus
5 minutes resting

Cooking time:
45 minutes

Prep ahead:
- Roast sweet potatoes
- Cook lentils
- Make pesto
- Roast macadamias

8 small orange sweet potatoes (about 90 g each)
½ cup (110 g) puy-style or green lentils (see Glossary)
300 g rump steak
freshly ground black pepper
1 tablespoon extra virgin coconut oil
2 tablespoons shredded coconut
1 tablespoon tamari (see Glossary)
1 tablespoon freshly squeezed lime juice

Coriander pesto
1 cup (20 g) coriander leaves
1 cup (40 g) shredded kale
3 tablespoons sunflower seeds
2 tablespoons roasted macadamias
1 garlic clove, chopped
sea salt
½ cup (125 ml) olive oil

This is a beautiful meal and a perfect example of how delicious real food is! All of these recipes have been designed to highlight how yum scrum a plant-based diet can be, so you can easily remove the steak and add sauteed greens if you're vegetarian or simply feel like a meat-free day.

Preheat the oven to 200°C (400°F) and line a baking tray with baking paper.

Place the whole unpeeled sweet potatoes on the prepared tray and bake for 40 minutes or until very soft.

Meanwhile, cook the lentils in a saucepan of boiling water for 20 minutes or until just soft. Rinse and drain well.

To make the pesto, place the coriander, kale, sunflower seeds, macadamias, garlic and a good pinch of salt in a food processor (the salt helps to stop the pesto from turning black) and process until smooth. With the motor running, add the olive oil and process until well combined.

Pat the steak dry and season both sides with pepper. Heat the coconut oil in a frying pan over medium–high heat, add the steak and cook for 5 minutes on each side or until done to your liking. Remove and rest, covered, for 5 minutes. Cut the rested steak into thin slices and place in a bowl. Add the coconut, tamari and lime juice and mix to combine.

Spread the pesto over four plates, top with the sweet potatoes and coconut steak salad, then scatter over the lentils.

CAULIFLOWER & BROCCOLI FRIED RICE

Serves:
4

Preparation time:
20 minutes

Cooking time:
15 minutes

Prep ahead:
- Process cauliflower and broccoli
- Chop vegetables

½ small head cauliflower (about 350 g), cut into florets
250 g broccoli, cut into florets
2 tablespoons extra virgin coconut oil
2 eggs, lightly beaten
2 spring onions, thinly sliced
1 tablespoon grated ginger
1 celery stalk, finely chopped
1 red capsicum, seeded and diced
½ cup (100 g) fresh corn kernels
½ cup (60 g) frozen peas, thawed
½ cup (125 ml) good-quality chicken stock
1 tablespoon tamari (see Glossary)
2 tablespoons chopped coriander leaves
2 tablespoons toasted sesame seeds
few drops of sesame oil

A healthy alternative to traditional fried rice, this is a great way to add a serving from the nutrient-rich brassica family. Don't worry if you don't have a food processor; you can still make this recipe – you just need a sharp knife. Shave the cauliflower and broccoli, then chop it into rice-sized pieces.

Put the cauliflower and broccoli in a food processor and pulse to small pieces roughly the size of rice.

Melt 1 teaspoon of the coconut oil in a wok over medium heat, add the egg and swirl to coat the base. Cook for 2 minutes or until the edge is set, then flip it over and cook the other side.

Remove the omelette from the wok and allow to cool slightly, then cut into thin strips.

Melt the remaining oil in the wok, add the spring onion, ginger, celery and capsicum and stir-fry for 5 minutes or until the vegetables are just soft.

Add the cauliflower and broccoli mixture, corn, peas, stock and tamari and stir-fry for 5 minutes or until the cauliflower is just tender.

Remove from the heat and fold through the omelette strips, coriander and toasted sesame seeds. Add a few drops of sesame oil and serve hot.

CHAPTER THREE

SNACKS

BROCCOLI & SPINACH FLATBREAD

Serves:
4

Preparation time:
15 minutes

Cooking time:
25 minutes

Prep ahead:
• Chop broccoli

1 head broccoli (320 g), roughly chopped
1 cup (30 g) baby spinach leaves
1 cup (120 g) almond meal
3 eggs, lightly beaten
¼ teaspoon ground turmeric (see Glossary)
sea salt and freshly ground black pepper

This is a snack option with serious nutritional value. Who would have thought you could sneak broccoli into a flatbread alternative! Broccoli is a great source of sulphoraphane, which is a powerful antioxidant – your liver loves it. The broccoli can be replaced with cauliflower, if preferred.

Preheat the oven to 200°C (400°F) and line a baking tray with baking paper.

Place the broccoli and spinach in a food processor and process until the mixture is very finely chopped – if it is too coarse it won't hold together.

Transfer the mixture to a bowl, add the almond meal, egg and turmeric, and season with salt and pepper.

Tip the mixture onto the prepared baking tray and spread out to a large rectangle about 5 mm thick. Bake for 20–25 minutes until dry. Remove from the oven and cool completely before cutting into smaller cracker-sized rectangles, then serve scattered with a pinch or two of sea salt, if you like.

If not eating them straight away, keep the flatbread in an airtight container in the refrigerator for up to 3 days.

SEED CRACKERS WITH CREAMY SALMON DIP

Serves:
4

Preparation time:
30 minutes, plus overnight soaking

Cooking time:
1 hour 50 minutes

Prep ahead:
- Soak and cook beans

½ cup (75 g) linseeds
½ cup (75 g) pumpkin seeds
½ cup (75 g) sesame seeds
½ cup (70 g) chia seeds
½ cup (60 g) almond meal
finely grated zest of 1 lemon
sea salt
1 tablespoon hulled tahini, at room temperature (see Glossary)
¾ cup (185 ml) water

Salmon dip
½ cup (110 g) dried cannellini beans
150 g hot-smoked salmon or trout (see Glossary)
⅓ cup (80 ml) extra virgin olive oil
2 tablespoons freshly squeezed lemon juice
2 tablespoons chopped dill
sea salt and freshly ground black pepper

Crackers are generally a pantry staple, but often contain preservatives, highly refined flours and poor-quality fats. This delicious, nutrient-dense recipe literally just uses nuts and seeds, which are rich in minerals that help us relax. To save time, double the quantity and store the leftovers in an airtight container for up to 2 weeks.

To make the salmon dip, place the cannellini beans in a bowl, cover with cold water and leave to soak overnight. The next morning, rinse and drain well.

Cook the beans in a saucepan of boiling water for 1 hour or until very soft. Drain well. Transfer to a food processor, add the salmon or trout, olive oil, lemon juice and dill and process until smooth and creamy. Season to taste with salt and pepper, then set aside to cool while you make the crackers.

Preheat the oven to 160°C (315°F) and line a large baking tray with baking paper.

Place the seeds, almond meal, lemon zest and salt in a bowl, stir in the tahini and water and mix to combine. Spread the mixture out on the prepared baking tray to a 5 mm thickness and bake for 40 minutes, then carefully turn over and cook the other side for a further 10 minutes or until the cracker is crisp and dry. Remove and cool completely, then break into large pieces and serve with the salmon dip.

CARROT CAKE BLISS BALLS

Makes:
16

Preparation time:
15 minutes, plus
30 minutes chilling

Prep ahead:
- Grate carrot
- Chop dates

1 cup (180 g) grated carrot (from 1 large carrot)
1 cup (190 g) medjool dates, pitted and roughly chopped (see Glossary)
3 tablespoons pumpkin seeds
3 tablespoons sunflower seeds
¼ teaspoon ground turmeric (see Glossary)
½ teaspoon ground cinnamon
2 teaspoons extra virgin coconut oil, melted
sea salt
3 tablespoons desiccated coconut

I so often find that people's energy and nourishment are let down by their snack choices. Not with these! These bliss balls have the delicious flavour of carrot cake, without the fuss of baking, which also means all of the nutrients are retained. They will keep for 5 days in the refrigerator, so they're a great one to prepare in advance. If you would like the balls to be firmer, double the quantity of coconut oil.

Place the carrot, dates, seeds, spices, coconut oil and a pinch of salt in a food processor and process until combined and starting to form a ball that comes away from the side.

Shape tablespoons of the mixture into balls and indent the top using your thumb, then press the tops in the coconut. Chill for 30 minutes or until firm.

CAULIFLOWER & CUMIN BITES WITH TAHINI DIP

Serves:
4–6

Preparation time:
15 minutes

Cooking time:
45 minutes

Prep ahead:
- Cut cauliflower into florets
- Make tahini dip

1 head cauliflower, cut into florets
1 teaspoon sea salt
2 teaspoons cumin seeds
2 tablespoons olive oil

Tahini dip
½ cup (140 g) hulled tahini, at room temperature (see Glossary)
½ cup (125 ml) water, plus extra if needed
2 tablespoons freshly squeezed lemon juice
¼ teaspoon ground turmeric (see Glossary)
1 garlic clove, crushed
sea salt

Nutrient-dense and great for digestion, this recipe can also easily be made into a delicious salad. Scatter the cauliflower over mixed greens and chargrilled beef, sprinkle with seeds and drizzle with the thinned tahini dip. The tahini dip is also delicious served with the Rainbow rice patties on page 142 or the Salmon and quinoa fish cakes on page 148.

Preheat the oven to 200°C (400°F) and line a large baking tray with baking paper.

Place the cauliflower florets on the prepared tray, sprinkle with the salt and cumin seeds and drizzle with the olive oil. Bake for 45 minutes or until the cauliflower has dried out and blackened in parts, but is not too dark.

Meanwhile, to make the tahini dip, place all the ingredients in a bowl and mix well to combine. Add a little more water if you would like it to be a thinner consistency.

Serve the cauliflower with the dip on the side.

PUMPKIN WEDGES WITH DUKKAH CRUST

Serves:
4–6

Preparation time:
25 minutes

Cooking time:
1 hour

Prep ahead:
- Make dukkah
- Roast pumpkin

1 kg kent pumpkin, skin on, cut into 1.5 cm thin wedges
2 tablespoons olive oil

Dukkah
½ cup (70 g) pistachios or hazelnuts
⅔ cup (100 g) sesame seeds
2 tablespoons coriander seeds
1 teaspoon cumin seeds
1 teaspoon freshly ground black pepper
½ teaspoon ground cinnamon
1 teaspoon sea salt

Pumpkin contains an array of vitamins and minerals, including beta-carotene which is wonderful for immunity, energy and vision. These wedges are delicious served on their own or as an accompaniment to roast chicken. This dukkah recipe makes 1.5 cups (200 g) of dukkah, which is more than you need here; the leftovers will keep in an airtight container for up to 3 months. Sprinkle it over grilled meat, fish, poultry or other roast vegetables, or over soups or hummus. You could also enjoy it simply with flatbread (see the Broccoli and spinach flatbread on page 158).

Preheat the oven to 200°C (400°F) and line a large baking tray with baking paper.

To make the dukkah, place the nuts and seeds in a frying pan and toast over low–medium heat for 10–15 minutes until they start to brown and become fragrant.

Transfer the mixture to a food processor, add the remaining ingredients and pulse briefly until just combined (you don't want to overdo it otherwise it will form a paste).

Place the pumpkin wedges in a large bowl with the olive oil and ½ cup (70 g) of the dukkah mixture and toss until the pumpkin is well coated.

Arrange the pumpkin wedges on the prepared tray and bake for 45 minutes or until cooked through and crisp around the edges.

SIX QUICK SNACKS

Each recipe serves 4

VEG STICKS & AVOCADO WRAPPED IN LETTUCE

- 2 cm piece ginger, peeled and cut into matchsticks
- 1 tablespoon apple cider vinegar
- 16 soft lettuce leaves (such as green oak), trimmed
- ½ carrot, cut into matchsticks
- ½ red capsicum, seeded and cut into matchsticks
- ½ celery stalk, cut into matchsticks
- 1 avocado, sliced
- 2 tablespoons tamari (see Glossary)
- drop of sesame oil

Combine the ginger and vinegar in a small bowl and leave to soak for 5 minutes, then drain.

Meanwhile, lay the lettuce leaves on a chopping board. Overlap two lettuce leaves on top of each other to form 8 circles, then divide the carrot, capsicum, celery, avocado and ginger between them. Roll the lettuce leaves up to enclose the filling.

Combine the tamari and sesame oil in a small bowl and serve with the lettuce rolls.

PEPPER & HONEY NUTS

200 g mixed nuts
100 g mixed seeds
1 tablespoon olive oil
½ teaspoon freshly ground black pepper
½ teaspoon sea salt
½ teaspoon ground turmeric (see Glossary)
2 teaspoons raw honey

Preheat the oven to 180°C (350°F). Line a baking tray with baking paper.

Place the nuts and seeds in a bowl and add the olive oil, pepper, salt, turmeric and honey. Mix to combine. Spread over the prepared tray and bake for 10–15 minutes until crisp and golden.

RAINBOW AVOCADO BOATS

1 tablespoon apple cider vinegar
1 teaspoon tamari (see Glossary)
1 tablespoon extra virgin olive oil
1 small carrot, grated
handful of snow pea sprouts
2 avocados, halved and stones removed
⅓ cup (50 g) red cabbage sauerkraut, drained on paper towel
1 tablespoon sunflower seeds

Combine the apple cider vinegar, tamari and olive oil.

Place the carrot and sprouts in a bowl and mix to combine. Divide between the avocado halves.

Top each avocado half with sauerkraut and sunflower seeds, then drizzle with the dressing.

>>

ROASTED SWEET POTATO BITES

350 g sweet potato, peeled and
cut into 1 cm slices
¼ cup (60 g) good-quality hummus
handful of alfalfa sprouts
¼ cup (35 g) sauerkraut
sweet paprika, to sprinkle

Preheat the oven to 200°C (400°F) and line a baking tray with baking paper.

Arrange the sweet potato discs on the prepared tray and bake for 30 minutes or until tender. Remove and allow to cool.

Top the discs with hummus, alfalfa sprouts and sauerkraut, and finish with a sprinkle of paprika.

IRANIAN FIGS

1 orange, segmented
½ teaspoon vanilla extract
1 teaspoon raw honey
4 fresh figs, halved lengthways
2 tablespoons hulled tahini, at room temperature (see Glossary)

Place the orange segments, vanilla and honey in a bowl and gently mix to combine.

Arrange the figs on a plate. Top with the tahini and the orange segments.

TAMARI & GINGER NUTS & SEEDS

250 g mixed nuts
100 g mixed seeds
1 tablespoon finely grated ginger
2 tablespoons tamari (see Glossary)
1 teaspoon olive oil

Preheat the oven to 180°C (350°F) and line a baking tray with baking paper.

Place the nuts and seeds in a bowl, add the ginger, tamari and olive oil and mix to combine. Spread over the prepared tray and bake for 10–15 minutes until crisp and golden.

SWEET HALVA DIP WITH FRESH FRUIT

Serves:
4

Preparation time:
10 minutes

½ cup (140 g) hulled tahini, at room temperature (see Glossary)
1 teaspoon ground cinnamon
1 teaspoon raw honey
1 teaspoon vanilla powder or extract
1 tablespoon chopped raw pistachios
200 g mixed strawberries, raspberries and/or blueberries
2 bananas, sliced

The halva dip can be used as a sweet spread on gluten-free crackers (such as the Seed crackers on page 161), or try serving it with the Flour-free banana pancakes on page 105.

Place the tahini, cinnamon, honey and vanilla in a bowl and mix until smooth and well combined.

Transfer the dip to a serving bowl and top with the chopped pistachios.

Serve with fresh berries and banana slices, for dipping.

ROSE-INFUSED ALMONDS & CURRANTS

Serves:
4

Preparation time:
10 minutes, plus overnight soaking

Prep ahead:
• Soak almonds and currants

1 cup (160 g) raw almonds
3 tablespoons currants
½ teaspoon rosewater
1 teaspoon vanilla extract
finely grated zest of 1 orange

This mix is also a perfect addition for breakfast – simply roughly chop and sprinkle over fresh fruit or porridge. Sometimes people overeat due to a lack of taste stimulation; if you relate to that, you must try these! The rosewater adds a fragrant Middle Eastern note but it can be left out if preferred, as the orange and vanilla still offer plenty of lovely flavours.

Place the almonds and currants in a bowl, cover with cold water and leave to soak overnight. The next morning, rinse, drain well, then pat dry.

Place the almonds and currants in a bowl, add the rosewater, vanilla and orange zest and mix well to combine. Store in an airtight container in the refrigerator for up to 5 days.

OLIVE & ROSEMARY PAN BREAD

Makes:
1 loaf

Preparation time:
15 minutes, plus
15 minutes standing

Cooking time:
30 minutes

3 eggs, lightly beaten
3 tablespoons chia seeds
2 teaspoons apple cider vinegar
2 tablespoons olive oil
1 ½ teaspoons bicarbonate of soda
1 teaspoon baking powder
2 tablespoons pumpkin seeds
2 tablespoons sunflower seeds
2 cups (240 g) almond meal
2 tablespoons chopped rosemary leaves, plus extra leaves to garnish
sea salt
100 g kalamata olives, pitted and sliced

A gluten-free version of focaccia! If you typically feel tired after eating bread, this is a great option for you. To make a sweet version of this loaf, replace the olives with dates and omit the rosemary and salt. Store the bread in an airtight container in the refrigerator for up to 3 days.

Preheat the oven to 200° (400°F). Place a 24 cm × 8 cm loaf tin in the oven for 20 minutes to warm.

Place the egg, chia seeds, vinegar, olive oil, bicarbonate of soda, baking powder, seeds, almond meal, rosemary, salt to taste and most of the olives (reserve a few to decorate the top) in a bowl and mix to combine. Set aside for 15 minutes.

Place a piece of baking paper on a work surface. Shape the dough into a loaf slightly smaller than the tin and top it with the remaining olives, extra rosemary and a little more salt.

Remove the loaf tin from the oven and very carefully transfer the paper topped with the uncooked loaf into the tin. Return the tin to the oven and bake the loaf for 30 minutes or until firm and a skewer inserted in the centre comes out clean.

SPICED CHICKPEA, CASHEW & KALE MIX

Serves:
4

Preparation time:
15 minutes, plus overnight soaking

Cooking time:
1 hour 15 minutes

Prep ahead:
- Soak and cook chickpeas

½ cup (100 g) dried chickpeas
1 teaspoon ground turmeric (see Glossary)
1 teaspoon ground cumin
1 teaspoon sweet paprika
sea salt
1 tablespoon extra virgin coconut oil, melted
½ cup (75 g) raw cashews
½ cup (40 g) shredded coconut
1 cup (40 g) shredded kale

A nutritious alternative to commercial versions of roasted mixed nuts, these are particularly great as an afternoon snack to avoid those 3 pm sugar cravings. The good fats and protein will keep you fuelled until dinner.

Place the chickpeas in a bowl, cover with cold water and leave to soak overnight. The next morning, rinse and drain well.

Preheat the oven to 180°C (350°F) and line a baking tray with baking paper.

Place the chickpeas in a bowl, add the spices, salt to taste and half the coconut oil then toss to coat. Spread the chickpeas out on the baking tray and bake for 40 minutes.

Add the cashews, coconut, kale and remaining coconut oil and bake for a further 5 minutes or until the kale is crisp. Remove and allow to cool. Store in an airtight container for up to 5 days.

SALMON SASHIMI WITH TAMARI GINGER DIPPING SAUCE

Serves:
4

Preparation time:
15 minutes

Prep ahead:
- Slice salmon
- Make dipping sauce

400 g sashimi-grade salmon
1 avocado, cut into 5 mm slices
1 punnet (125 g) mustard cress

Dipping sauce
1 tablespoon finely grated ginger
½ teaspoon sesame oil
3 tablespoons tamari (see Glossary)
1 tablespoon apple cider vinegar

When we think of snack foods we tend to think of crackers, nuts, and perhaps a muesli bar – but you can make a snack with anything that's in your fridge. Snacks are another great opportunity to nourish your body and to sustain your energy levels throughout the day and this one will give you a boost. It's important to use sashimi-grade salmon, if you're having this raw.

Using a very sharp fillet knife, remove the skin from the salmon, then cut it into 5 mm slices.

Arrange the avocado on a serving plate with the mustard cress and salmon slices.

To make the dipping sauce, combine all the ingredients in a small bowl.

Serve the sashimi salmon and avocado salad with the dipping sauce on the side.

ROAST CAULIFLOWER DIP WITH ENDIVE BOATS, TOMATOES & CARROTS

Serves:
4

Preparation time:
20 minutes, plus overnight soaking

Cooking time:
1 hour

Prep ahead:
- Soak and cook cannellini beans

½ cup (110 g) dried cannellini beans
½ head cauliflower, cut into florets
1 garlic clove, chopped
⅓ cup (80 ml) extra virgin olive oil
3 tablespoons water
sea salt
2 fresh endives, leaves separated
2 carrots, peeled, quartered lengthways and thickly sliced
250 g cherry tomatoes, halved

The *Brassica* family of vegetables remains one of my all-time favourite vegetable varieties due to their anti-cancer properties and liver detoxification support. The cannellini beans give this dip the substance to see you through to your next main meal and pack a mighty nutritional punch to boot.

Place the cannellini beans in a bowl, cover with cold water and leave to soak overnight. The next morning, rinse and drain well.

Cook the cannellini beans in a saucepan of boiling water for 1 hour or until very soft. Rinse and drain well.

Preheat the oven to 200°C (400°F) and line a baking tray with baking paper.

While the beans are cooking, place the cauliflower on the prepared tray and roast for 40 minutes or until soft and golden.

Place the roast cauliflower, cooked beans, garlic, olive oil, water and a pinch of salt in a food processor and process until very smooth and creamy.

Transfer the dip to a bowl and serve with the endive leaves, carrot sticks and cherry tomatoes.

BEETROOT, CACAO & CHIA MUFFINS

Makes:
8

Preparation time:
15 minutes

Cooking time:
25 minutes

Prep ahead:
• Grate beetroot

1 cup (180 g) finely grated beetroot (from 1 beetroot)
1 cup (120 g) almond meal
½ cup (40 g) psyllium husks
3 tablespoons raw cacao powder, plus extra for dusting
3 tablespoons chia seeds
1 teaspoon baking powder
2 eggs, lightly beaten
3 tablespoons pure maple syrup
3 tablespoons extra virgin coconut oil, melted
2 ripe bananas, mashed (you'll need ¾ cup)

Betacyanin, the pigment that gives beetroot its colour, is a powerful antioxidant used by the body to counteract free radicals, reducing cell damage and inflammation, and slowing the aging process from the inside out. Beetroot also contains substances that enhance oxygenation, essential for great energy. These are moist muffins, so they freeze well. Store in airtight containers and thaw as needed.

Preheat the oven to 180°C (350°F) and line 8 x ⅓ cup (80 ml) muffin holes with paper cases.

Place the beetroot, almond meal, psyllium husks, cacao powder, chia seeds and baking powder in a bowl and mix well to combine.

Whisk together the egg, maple syrup, coconut oil and banana, then stir this mixture into the dry ingredients until just combined.

Spoon the batter evenly into the paper cases. Bake for 25 minutes or until risen and cooked through. Cool in the tin for a few minutes, then transfer to a wire rack to cool a little more. Serve warm, dusted lightly with extra cacao powder.

CHAPTER FOUR

DINNER

PUMPKIN, MISO & LEEK RISOTTO WITH TAMARI SEED SPRINKLE

Serves:
4

Preparation time:
25 minutes

Cooking time:
45 minutes

Prep ahead:
- Grate pumpkin
- Make tamari seed sprinkle

1.5 litres good-quality chicken stock (or vegetable stock if you want to make this vegetarian)
3 tablespoons olive oil
1 leek, white part only, thinly sliced
2 cups (400 g) arborio rice
3 cups (360 g) grated pumpkin
2 teaspoons shiro miso or genmai miso (see Glossary)
2 tablespoons ghee (see Glossary)
3 tablespoons picked flat-leaf parsley, dill and oregano

Tamari seed sprinkle

½ cup (75 g) mixed seeds (such as pumpkin, sunflower and sesame seeds)
1 tablespoon finely grated ginger
2 tablespoons tamari (see Glossary)

Sometimes supporting better energy means embracing versatile meals that can be adapted and used again. This risotto is a fantastic example. Leftovers can be taken for lunch the next day or you can turn them into arancini. Simply shape the cooled risotto into balls, coat them with seeds and brush with oil, then bake in the oven until golden.

Pour the stock into a saucepan and bring to the boil. Reduce the heat and keep it at a low simmer.

Heat the olive oil in a large deep frying pan over medium heat, add the leek and cook for 10 minutes or until soft and golden.

Add the rice and cook for 2 minutes or until it is coated with the oil. Add the pumpkin and cook, stirring, for 5 minutes or until the pumpkin starts to soften.

Add 1 cup (250 ml) of the simmering stock to the rice and cook, stirring, until all of the stock has been absorbed. Continue adding the stock 1 cup at a time, allowing each cup to be fully absorbed before adding the next. Once all the stock has been absorbed and the rice is soft and creamy, stir in the miso and ghee. Remove from the heat and cover while you prepare the seeds.

To make the tamari seed sprinkle, place the seeds in a small frying pan over medium heat and cook until they start to colour. Add the ginger and tamari, then reduce the heat to low and cook for 3–5 minutes until the tamari has been absorbed by the seeds.

Ladle the risotto into shallow bowls and top with the tamari seeds and herbs.

LEMONGRASS BEEF BROTH WITH ZUCCHINI & CARROT NOODLES

Serves:
4

Preparation time:
20 minutes

Cooking time:
35 minutes

Prep ahead:
- Slice spring onion and ginger
- Slice beef

1 × 400 ml tin coconut cream
1 litre good-quality chicken stock
2 lemongrass stalks, white part only, pounded until flat
8 kaffir lime leaves, torn (see Glossary)
4 spring onions, sliced
1 tablespoon sliced ginger
1 tablespoon freshly squeezed lime juice
1 tablespoon tamari (see Glossary)
120 g baby corn
200 g baby bok choy, leaves separated, larger leaves sliced
2 zucchini, spiralised into thin noodles
2 carrots, spiralised into thin noodles
100 g bean sprouts, trimmed
300 g scotch fillet steak, trimmed and very thinly sliced
1 long red chilli, very thinly sliced
handful of coriander leaves

I love the combination of lemongrass, kaffir lime leaves and coriander in this beautiful broth. The nutrient density here is increased by using zucchini noodles, as opposed to rice or egg noodles. The beef is cooked in the hot liquid so it's important it is very thinly sliced.

Combine the coconut cream, stock, lemongrass, kaffir lime leaves, spring onion and ginger in a large saucepan and bring to the boil over high heat. Reduce the heat and simmer for 30 minutes, then strain, discarding the aromatics.

Return the liquid to the pan, stir in the lime juice and tamari and bring back to a simmer. Add the baby corn and bok choy, and remove the pan from the heat.

Divide the zucchini and carrot noodles among four large heatproof bowls, top with the bean sprouts and beef and ladle the hot stock over the top. Finish with the sliced chilli and coriander and serve.

HEALING CHICKEN SOUP

Serves:
6

Preparation time:
15 minutes

Cooking time:
4 hours

Prep ahead:
- Rinse chicken and pat dry
- Grate turmeric
- Chop vegetables

1 × 1.6 kg organic chicken
1 garlic head, halved crossways
1 tablespoon sliced ginger
1 tablespoon grated fresh red turmeric (see Glossary)
4 star anise
6 black peppercorns
1 brown onion, chopped
2 carrots, chopped
3 celery stalks, sliced
3 litres water
200 g cauliflower, cut into florets
2 parsnips, peeled and sliced
2 zucchini, sliced into rounds
200 g broccoli, cut into florets, stalks chopped
1 tablespoon tamari (see Glossary)
flat-leaf parsley leaves, to serve

This is the perfect recipe to either help prevent or treat winter ailments – a great go-to for when you don't quite feel yourself. For an additional green boost, add finely chopped silverbeet or kale. This soup can also be frozen and reheated as required.

Rinse the chicken and pat dry with paper towel. Place the chicken in a large stockpot and add the garlic, ginger, turmeric, star anise, peppercorns, onion, carrot and celery.

Pour in the water and bring to the boil, then reduce the heat and simmer for 3 hours or until the chicken is falling off the bone. Remove the chicken bones and discard. Skim off any foam that rises to the surface of the soup during this time.

Add the remaining ingredients and cook for 1 hour or until the vegetables are tender. Ladle generously into bowls and serve scattered with parsley.

WARMING DHAL

Serves:
4

Preparation time:
15 minutes, plus standing time

Cooking time:
25 minutes

Prep ahead:
- Grate sweet potato and cauliflower
- Chop tomatoes
- Shred kale

2 cups (420 g) red lentils, rinsed
1 teaspoon ground turmeric (see Glossary)
1 cinnamon stick
1 tablespoon grated ginger
1 long red chilli, halved (optional)
1 brown onion, chopped
2 cups (240 g) grated sweet potato
2 cups (115 g) grated cauliflower
300 g ripe tomatoes, chopped
1 litre water
2 teaspoons brown mustard seeds
sea salt
2 cups (80 g) finely shredded kale
juice of 1 lemon
2 tablespoons ghee (see Glossary)
2 teaspoons cumin seeds

Dhal nurtures the soul. It is a wonderfully restorative meal for those who bloat easily or feel depleted. The warming spices can also aid digestion. Make double and freeze the other portion in individual servings – particularly good for avoiding those Friday-night takeaways.

Put the lentils, turmeric, cinnamon stick, ginger, chilli, if using, onion, sweet potato, cauliflower, tomato, water, half the mustard seeds and a generous pinch of salt in a large saucepan. Bring to the boil over high heat, then reduce the heat and simmer for 20 minutes or until thick and creamy. Remove from the heat and stir through the kale and lemon juice, then cover and set aside for 5 minutes.

Melt the ghee in a small frying pan over medium heat, add the cumin seeds and remaining mustard seeds and cook for 2 minutes or until the mustard seeds start to pop.

Ladle the dhal into four bowls, top with the cumin and mustard seeds and serve.

BROWN RICE KITCHARI

Serves:
6

Preparation time:
20 minutes, plus overnight soaking

Cooking time:
40 minutes

Prep ahead:
- Soak rice and dhal
- Chop or grate broccoli
- Remove kernels from corn cob
- Grate zucchini and carrot

½ cup (110 g) medium-grain brown rice
½ cup (105 g) split mung dhal (see Glossary)
2 tablespoons extra virgin coconut oil
⅓ cup (80 g) ghee (see Glossary)
1 brown onion, finely chopped
1 tablespoon grated ginger
1 tablespoon black mustard seeds
1 tablespoon black sesame seeds
2 teaspoons cumin seeds
2 teaspoons grated fresh turmeric or ½ teaspoon ground turmeric (see Glossary)
1 cup (85 g) grated or finely chopped broccoli
1 corn cob, kernels removed (1 cup/200 g)
2 zucchini, grated
2 carrots, grated
1.25 litres boiling water
sea salt
freshly squeezed lemon juice, to taste
chopped coriander leaves, to serve

Kitchari is considered a food medicine, and this nourishing meal is very gentle on the digestive system. It is delicious reheated and can also be served as an accompaniment to other dishes. Basmati rice can be used instead of the brown rice and this will reduce the cooking time.

Wash the rice and dhal, then place in a bowl, cover with cold water and leave to soak overnight. The next morning, rinse and drain well.

Melt the coconut oil and half the ghee in a large deep frying pan over medium heat, add the onion and ginger and cook for 10 minutes or until the onion is soft and golden – don't rush this step as the gently cooked onion adds a lovely sweetness to the dish.

Add the seeds and turmeric and cook for 1 minute or until the seeds start to pop. Add the broccoli, corn, zucchini and carrot and stir until the vegetables are well coated in the spices.

Stir in the drained rice and dhal, then add the boiling water and a good pinch of salt. Bring to the boil over high heat and boil for 20 minutes or until tunnels form in the rice and most of the liquid has been absorbed. Reduce the heat to low, cover and cook gently for 5 minutes or until the rice is soft and there is no liquid left in the pan.

Stir through the remaining ghee, then finish with a good squeeze of lemon and a scattering of coriander.

SMOKY CHICKEN WITH REFRIED BLACK BEANS

Serves:
4

Preparation time:
15 minutes, plus overnight soaking and marinating time

Cooking time:
50 minutes

Prep ahead:
- Soak and cook beans
- Marinate chicken

1 cup (200 g) dried black beans
500 g organic chicken thigh fillets, skin removed
1 teaspoon finely grated lime zest
2 tablespoons freshly squeezed lime juice
1 teaspoon sweet smoked paprika
sea salt
3 tablespoons olive oil
3 spring onions, thinly sliced
1 teaspoon ground cumin
1 corn cob, kernels removed (1 cup/200 g)
250 g cherry tomatoes, halved
100 g baby spinach leaves
3 tablespoons coriander leaves
lime wedges, to serve (optional)

To make this meal vegetarian, omit the chicken and add a fried or poached egg and half an avocado seasoned with lemon juice, salt and pepper. This is best cooked on the barbecue (it adds to the smoky flavour), but can also be cooked in a frying pan.

Place the beans in a bowl, cover with cold water and leave to soak overnight. The next morning, rinse and drain well. Cook the beans in boiling water for 30 minutes or until just soft.

Place the chicken fillets in a large non-metallic dish, add the combined lime zest, lime juice, paprika and salt to taste. Gently toss to coat well, then leave to marinate for as long as time permits.

Heat half the olive oil on a barbecue flat plate over medium–high heat, add the chicken and cook for 5 minutes on each side or until tender and cooked through. Set aside for 5 minutes while you finish the beans.

Heat the remaining oil on the barbecue flat plate, add the spring onion and cook for 5 minutes or until soft. Add the black beans, cumin, corn and tomato and cook for a further 5 minutes or until heated through.

Transfer the bean mixture to a bowl and stir through the spinach and coriander.

Serve the beans and chicken thighs with lime wedges on the side, if using.

CURRY ROASTED VEGETABLES WITH COCONUT EGGS

Serves:
4

Preparation time:
30 minutes

Cooking time:
40 minutes

Prep ahead:
- Cut up vegetables
- Hard-boil eggs

3 large carrots, cut into batons
3 parsnips, peeled and cut into batons
400 g cauliflower, cut into florets
500 g kent pumpkin, seeded, skin left on, cut into thin wedges
1 red onion, cut into wedges
1 red capsicum, seeded and cut into strips
3 tablespoons extra virgin coconut oil, melted
3 tablespoons freshly squeezed lime juice
1 tablespoon black mustard seeds
1 tablespoon curry powder
sea salt

Coconut eggs
1 garlic clove, chopped
2 tablespoons chopped ginger
1 cup (40 g) coriander roots, stems and leaves
1 × 270 ml tin coconut cream
8 hard-boiled eggs, peeled

This is bliss to me: the perfect meal after a busy day. Eggs and vegetables are simple but often just what is required. Parsnip is a winter veg, so if you want to cook this dish in the warmer months, simply leave the parsnip out or replace it with a seasonal vegetable of your choice.

Preheat the oven to 200°C (400°F). Place the vegetables in a large bowl, add the coconut oil, lime juice, mustard seeds and curry powder and toss well to coat. Season to taste with salt.

Transfer to a baking dish and bake for 40 minutes or until the vegetables are soft.

Meanwhile, to make the coconut eggs, place the garlic, ginger and coriander in a small food processor and process until smooth.

Transfer the coriander paste to a small saucepan, add the coconut cream and bring to the boil over medium heat. Reduce the heat, then add the eggs and simmer for 15 minutes.

Serve the roasted vegetables topped with the coconut eggs.

MOROCCAN LAMB SHANKS & VEG WITH CORIANDER OIL

Serves:
4

Preparation time:
20 minutes

Cooking time:
4 hours 15 minutes

Prep ahead:
- Chop vegetables
- Make coriander oil

2 tablespoons ground coriander
1 teaspoon sweet paprika
4 lamb shanks
2 tablespoons olive oil, plus extra if needed
1 red onion, chopped
400 ml tomato passata
2 cups (500 ml) good-quality chicken stock
2 tablespoons finely shredded preserved lemon rind or 1 tablespoon finely grated lemon zest
6 medjool dates, pitted and sliced (see Glossary)
½ cup (100 g) buckwheat (see Glossary)
500 g sweet potato, cut into 3 cm cubes
2 zucchini, cut into 2 cm slices
1 red capsicum, seeded and sliced
sea salt and freshly ground black pepper

Coriander oil

3 tablespoons chopped coriander leaves
3 tablespoons extra virgin olive oil

Lamb is a source of iron and zinc, both essential for energy and nutrients many people don't get enough of. Shanks are an excellent cut as you also get the benefits of what the bones infuse. Remember to always choose organic, pasture-raised meats. This is particularly yummy cooked in a slow cooker, if you have one. Slow cookers are a great addition to a busy world because you can throw everything in, leave it on the kitchen bench cooking overnight and wake up in the morning to a nourishing, home-cooked meal. If you use a slow cooker, just increase the cooking time to 6 hours.

Combine the ground coriander and paprika and rub them over the lamb shanks. Heat the olive oil in a 5 litre flameproof casserole dish over medium heat, add the lamb shanks and cook until browned on all sides. Remove the shanks from the dish.

Add the onion and another tablespoon of oil if necessary and cook for 10 minutes or until golden. Return the lamb shanks to the dish and add the tomato passata, stock, lemon and dates. Bring to the boil, then reduce the heat to low and cook, covered, for 3 hours.

Add the buckwheat, sweet potato, zucchini and capsicum and cook for a further 1 hour or until the vegetables are soft. Season to taste with salt and pepper.

To make the coriander oil, put the coriander and oil in a small food processor and process until smooth.

Serve the lamb shank stew drizzled with the coriander oil.

PAPRIKA STEAK WITH TOMATO BEANS & GARLIC GREENS

Serves:
4

Preparation time:
15 minutes, plus overnight soaking and 5 minutes resting

Cooking time:
50 minutes

Prep ahead:
• Soak beans

1 tablespoon smoked paprika
1 teaspoon ground cumin
sea salt
4 × 200 g sirloin steaks
1 tablespoon olive oil
150 g baby spinach leaves
3 cups (120 g) shredded kale
1 garlic clove, sliced
2 tablespoons ghee (see Glossary)

Tomato beans
1 cup (200 g) dried butter beans (see Glossary)
sea salt and freshly ground black pepper
1 tablespoon olive oil
1 brown onion, chopped
1 teaspoon dried oregano
400 ml tomato passata
1 cup (250 ml) water

If you're looking for a quick and delicious alternative to a takeaway, here it is. The tomato beans are also a great breakfast idea and the garlic greens can be used as a side with any other meal.

To make the tomato beans, place the butter beans in a bowl, cover with cold water and leave to soak overnight. The next morning, rinse and drain well.

Transfer the beans to a saucepan, cover with water and bring to the boil. Boil for 30 minutes or until just tender – do not overcook or the beans will fall apart when reheated in the sauce.

Meanwhile, heat the olive oil in a large deep frying pan over medium heat, add the onion and cook for 10 minutes or until golden. Add the oregano, tomato passata and water. Bring to the boil, then reduce the heat to low and simmer for 10 minutes or until the sauce has thickened. Add the butter beans to the sauce and simmer for a further 5 minutes.

Combine the paprika, cumin and salt to taste on a plate, then coat both sides of the steaks in the spice mix. Heat the oil in a frying pan over medium–high heat, add the steaks and cook for 3 minutes on each side or until cooked to your liking. Remove and rest for 5 minutes.

Add the spinach and kale to the pan you cooked the steak in, stir through the garlic and ghee and cook until the greens are soft and just wilted. Serve the steaks with the wilted greens and the tomato beans.

SPICED MIDDLE EASTERN LAMB WITH HUMMUS & BEETROOT SALAD

Serves:
4

Preparation time:
20 minutes

Cooking time:
10 minutes

Prep ahead:
- Prepare vegetables for beetroot salad

300 g lamb or beef mince
1 teaspoon grated ginger
1 teaspoon ground coriander
½ teaspoon ground allspice
½ teaspoon dried chilli flakes
sea salt
1 cup (200 g) good-quality hummus
coriander and mint leaves, to serve

Beetroot salad
1 small beetroot, peeled and grated
1 carrot, grated
3 red radishes, thinly sliced
¼ small red cabbage, finely shredded
2 tablespoons extra virgin olive oil
1 tablespoon freshly squeezed lemon juice
1 tablespoon balsamic vinegar
1 teaspoon pure maple syrup
sea salt and freshly ground black pepper

The slightly peppery flavour of the radish in this meal gives great variety for your tastebuds. Chickpeas, when soaked and cooked properly, can be wonderfully supportive of establishing good gut bacteria. Always choose organic, pasture-fed animal protein if you include it in your diet.

To make the beetroot salad, place the beetroot, carrot, radish and cabbage in a bowl. Whisk together the olive oil, lemon juice, balsamic vinegar and maple syrup and pour over the salad. Toss to combine and gently squeeze the vegetables so they are really well coated in the dressing. Season to taste with salt and pepper.

Heat a large deep frying pan over medium–high heat, add the mince, ginger and spices and cook, stirring, for 8–10 minutes or until the mince is browned, breaking up any lumps with the back of a wooden spoon. Season with salt.

Spread some hummus on a large serving plate or individual plates and top with the mince mixture. Scatter over the fresh herbs and serve with the beetroot salad.

PEPPER BEEF STIR-FRY

Serves:
4

Preparation time:
20 minutes

Cooking time:
20 minutes

Prep ahead:
- Slice steak
- Prepare vegetables

500 g rump steak, thinly sliced
1 teaspoon freshly ground black pepper
½ teaspoon five-spice powder
2 tablespoons extra virgin olive oil
2 tablespoons extra virgin coconut oil
1 brown onion, thinly sliced into rounds
1 tablespoon finely grated ginger
1 bunch asparagus (180 g), trimmed and cut into 5 cm lengths
150 g snow peas, trimmed
200 g broccolini, cut into 5 cm lengths, including stalks
75 g sugar snap peas
200 g baby choy sum, leaves separated
100 ml good-quality beef stock
1 tablespoon tamari (see Glossary)
1 teaspoon sesame oil
1 teaspoon pure maple syrup (optional)
micro rocket, to serve (optional)

Everyone needs a quick stir-fry recipe and this is one for the whole family. The beef contains iron and zinc, both of which are essential for great energy. Or you might like to substitute the beef for any other animal protein or your preferred vegetarian protein option. Make a bit extra and take your leftovers to work the following day.

Place the beef, pepper, five-spice powder and olive oil in a bowl and mix well to combine.

Melt half the coconut oil in a wok over high heat, add the beef and stir-fry for 5 minutes or until browned. Remove the beef from the wok.

Melt the remaining coconut oil in the wok, add the onion and stir-fry for 5 minutes or until soft and golden. Add the ginger and cook for 1 minute.

Stir in the vegetables, stock and tamari and stir-fry for 5 minutes or until the vegetables are bright green and tender.

Return the beef to the wok and cook for 3 minutes or until heated through. Remove from the heat and stir in the sesame oil and maple syrup, if using. Serve immediately, scattered with micro rocket, if using.

SLOW-SIMMERED BEEF, PARSNIP & LENTIL STEW

Serves:
4

Preparation time:
15 minutes, plus overnight soaking

Cooking time:
3 hours 45 minutes

Prep ahead:
- Soak lentils
- Peel onions and garlic cloves
- Make sweet potato mash

½ cup (110 g) black beluga, green or puy-style lentils (see Glossary)
2 tablespoons olive oil
700 g beef cheeks or chuck steak, cut into 3 cm chunks
8 baby onions or golden shallots, peeled
8 garlic cloves, peeled and left whole
4 large flat mushrooms, thickly sliced
1 cup (250 ml) good-quality beef stock
1 cinnamon stick
1 bay leaf
4 black peppercorns
2 cups (500 ml) water
4 small parsnips, peeled and halved lengthways
1 teaspoon fennel seeds
2 star anise
2 strips orange rind, plus extra finely grated zest to serve
1 cup (250 ml) freshly squeezed orange juice
1 tablespoon thinly sliced ginger
3 tablespoons chopped flat-leaf parsley
mashed sweet potato (see page 148), to serve

This is a wonderfully hearty winter stew. I add lentils to many meals to not only increase the fibre and protein content, but also because it means you can reduce the amount of meat used. You can cook this stew in a slow cooker if you prefer. Follow the recipe as directed then slow-cook for 4 hours or until the meat pulls apart easily.

Place the lentils in a bowl, cover with cold water and leave to soak overnight. The next morning, rinse and drain well.

Heat the olive oil in a 5 litre flameproof casserole dish over high heat, add the beef and cook until browned. Remove from the dish. Add the onions or shallots and cook until browned.

Return the meat to the dish and add the garlic, mushroom, stock, cinnamon stick, bay leaf, peppercorns, drained lentils and water. Bring to the boil, then reduce the heat to low and simmer, covered, for 2½ hours or until the beef pulls apart easily.

Add the parsnip, fennel seeds, star anise, orange rind, orange juice and ginger and cook, covered, for 1 hour or until the parsnip is soft.

Finish with a scattering of parsley and serve with sweet potato mash.

BEEF & SALSA VERDE GREENS

Serves:
4

Preparation time:
20 minutes, plus
5 minutes resting

Cooking time:
20 minutes

Prep ahead:
- Make the salsa verde

½ cup (100 g) red quinoa, rinsed well (see Glossary)
200 g baby chat potatoes
4 × 150 g sirloin steaks
sea salt and freshly ground black pepper
2 tablespoons olive oil
125 g mixed salad leaves

Salsa verde
2 tablespoons baby capers, rinsed
2 tablespoons green olives, pitted and chopped
1 garlic clove, chopped
2 tablespoons chopped flat-leaf parsley
2 tablespoons chopped basil
finely grated zest and juice of 1 lemon
3 tablespoons extra virgin olive oil
sea salt and freshly ground black pepper

Such a beautiful flavour combination. I've heard people exclaim that they could add this salsa verde to everything! What's more, you can have this meal on the table in a flash. Simply replace the beef with egg and additional greens if you're vegetarian.

Cook the quinoa in a saucepan of boiling water for 15 minutes or until the tails start to separate from the grain. Drain well and set aside.

Meanwhile, bring a medium saucepan of water to the boil, add the potatoes and cook over medium heat for 10 minutes or until just tender. Drain and slice, then cover and keep warm.

Pat the steaks dry with paper towel and season with salt and pepper. Heat the olive oil in a large deep frying pan over medium heat, add the steaks and cook for 5 minutes on each side or until cooked to your liking. Remove and rest for 5 minutes, then cut into thin slices.

To make the salsa verde, place all the ingredients in a small food processor and pulse to combine.

Pile the salad leaves into bowls, add the quinoa, potato, steak and half the salsa verde and gently toss to combine. Serve drizzled with the remaining salsa verde.

SLOW-COOKED LAMB SHOULDER WITH PEAR & BALSAMIC

Serves:
4

Preparation time:
15 minutes

Cooking time:
5 hours

Prep ahead:
- Stud lamb with garlic and rosemary
- Peel parsnips and cut vegetables

1 × 2 kg lamb shoulder, bone in
2 garlic cloves, sliced
2 rosemary sprigs, broken into short lengths
1 tablespoon freshly ground black pepper
2 tablespoons olive oil
2 small leeks, white and light green parts only, halved lengthways and washed
200 g baby chat potatoes, sliced
4 carrots, halved lengthways
4 parsnips, peeled and halved lengthways
2 pears, halved lengthways
1½ cups (375 ml) good-quality chicken stock
3 tablespoons balsamic vinegar
2 tablespoons flat-leaf parsley leaves, torn

Cooking meat with the bone in means you end up with beautifully tender meat that just falls apart. Couple this with pear and balsamic and your tastebuds will be very happy! Slow-cooked meat is wonderful when your digestive fire is compromised, as often happens when we spend out lives in our heads and over-thinking.

Preheat the oven to 160°C (315°F).

Pat the lamb shoulder dry with paper towel, then cut small slits into the meat. Press the garlic and rosemary into the slits. Sprinkle over the pepper.

Heat the olive oil in a large flameproof baking dish over medium heat, add the lamb shoulder and cook for 5 minutes or until browned on both sides. Remove from the heat.

Add the leek, potato, carrot, parsnip, pear, stock and balsamic vinegar. Cover with foil and roast for about 5 hours or until the lamb is very tender and shreds easily.

Remove the lamb from the oven and thickly shred the meat. Serve the shredded meat with the vegetables and pears and the broth spooned over the top. Finish with a scattering of torn parsley.

SIX QUICK VEG

Each recipe serves 4

TAMARI SNOW PEAS & BROCCOLINI

200 g snow peas
1 bunch broccolini, trimmed and halved
2 teaspoons extra virgin coconut oil
1 tablespoon tamari (see Glossary)
1 tablespoon toasted sesame seeds

Place the snow peas and broccolini in a deep frying pan and cover with water. Bring to the boil, then reduce the heat and simmer for 5 minutes or until the vegetables are bright green and tender. Remove from the pan and drain well.

Wipe and dry the frying pan, then place over medium heat. Add the coconut oil, then the snow peas, broccolini, tamari and sesame seeds. Cook for 3 minutes, stirring to coat the greens with the tamari and seeds.

JAPANESE SESAME GREEN BEANS

500 g green beans, trimmed and halved
1 tablespoon shiro miso (see Glossary)
3 tablespoons hot water
¼ cup (65 g) hulled tahini, at room temperature (see Glossary)
1 teaspoon sesame oil
2 tablespoons black sesame seeds

Steam the beans for 10 minutes or until bright green and tender, then transfer to a bowl.

Combine the miso and hot water and stir until smooth.

Place the tahini, sesame oil and miso mixture in a bowl and mix to combine. Pour the tahini dressing over the beans, toss gently to combine and sprinkle with the sesame seeds.

BAKED FENNEL WITH ALMOND CRUNCH

2 large fennel bulbs, trimmed and cut lengthways into 1 cm thick slices
3 tablespoons olive oil
100 g raw almonds, roughly chopped
¼ cup (40 g) sunflower seeds
handful of flat-leaf parsley leaves, roughly chopped
sea salt and freshly ground black pepper, to taste

Preheat the oven to 200°C (400°F) and line a baking tray with baking paper.

Arrange the fennel on the prepared tray and drizzle over half the olive oil. Bake for 30 minutes.

Meanwhile, place the almonds, sunflower seeds, parsley, salt, pepper and remaining olive oil in a food processor and process until roughly chopped.

Remove the fennel from the oven and top with the almond mixture. Return to the oven and bake for a further 15 minutes or until the fennel is tender and the topping is golden.

BEETROOT & RADISH CARPACCIO

2 tablespoons extra virgin olive oil
2 tablespoons apple cider vinegar
2 small beetroot, peeled and very thinly sliced with a mandoline
4 red radishes, very thinly sliced with a mandoline
sea salt and freshly ground black pepper

Whisk together the olive oil and vinegar.

Arrange the beetroot and radish on a serving plate, pour over the dressing and season with salt and pepper.

INDIAN COCONUT CARROTS

1 tablespoon extra virgin coconut oil
2 teaspoons brown mustard seeds
1 teaspoon ground turmeric (see Glossary)
3 carrots, grated
2 tablespoons desiccated coconut
2 tablespoons chopped coriander leaves

Heat the coconut oil in a large deep frying pan over medium heat. Add the mustard seeds and turmeric and cook until the mustard seeds just start to pop.

Add the carrot and coconut and cook for 5 minutes or until the carrot is just tender. Remove from the heat and stir through the coriander.

QUICK FRIED KALE & BABY SPINACH

2 tablespoons extra virgin olive oil
or extra virgin coconut oil
3 spring onions, sliced
100 g kale leaves, roughly torn
100 g baby spinach leaves
3 tablespoons pumpkin seeds
sea salt and freshly ground black pepper
2 tablespoons freshly squeezed lemon juice

Heat the oil in a large deep frying pan over medium heat. Add the spring onion and cook for 3 minutes or until golden.

Add the kale, spinach, pumpkin seeds and ½ teaspoon pepper and cook for 3–5 minutes or until the kale wilts. Drizzle with lemon juice and season with salt before serving.

ROAST LEMON CHICKEN

Serves:
4+

Preparation time:
15 minutes, plus
15 minutes resting

Cooking time:
1 hour

2 × 1.6 kg organic chickens, rinsed and patted dry with paper towel
2 brown onions, roughly chopped
2 carrots, roughly chopped
4 celery stalks, roughly chopped
2 bay leaves
2 garlic heads, halved crossways
2 lemons
6 rosemary sprigs
6 oregano sprigs
6 thyme sprigs
3 tablespoons ghee (see Glossary)
sea salt and freshly ground black pepper
3 tablespoons olive oil
green leaves, such as curly endive and baby rocket, to serve
thinly sliced Lebanese cucumber, to serve
thinly sliced lemon, to serve
Salsa verde (see page 213), to serve (optional)

This is a dish that keeps on giving. As a roast, it serves at least 4, with plenty of leftover meat for lunches or dinners during the week. In addition, the chicken carcasses can be used to make a delicious stock, perfect for warming soups.

Preheat the oven to 180°C (350°F).

Place the onion, carrot, celery, bay leaves and garlic in a baking dish large enough to hold both chickens. Put a lemon in each chicken cavity and pierce the lemon skin a couple of times with a fork. Divide the herbs between the two chickens and push them into the cavities with the lemon.

Carefully rub the ghee under the skin of the chickens, then season well and drizzle with the olive oil. Place the chickens on top of the vegetables and roast for 1 hour or until the juices run clear when a skewer is inserted into the thickest part of the leg.

Leave the chickens to rest for 15 minutes before carving. Transfer the roast vegetables and garlic into a large saucepan, ready to make the stock. Carve the chickens and serve with green leaves, cucumber and lemon, with salsa verde on the side, if you like.

Later, to make the stock, place the chicken carcasses in the saucepan with the vegetables and add water to generously cover. Bring to the boil, then reduce the heat and simmer for 3–4 hours or until the stock starts to become clear – skim off any foam that floats to the surface. Strain and pour the stock into an airtight container; it can be used straight away or frozen for later use in soups or stews.

CHICKEN TAGINE WITH YELLOW SPLIT PEA PUREE

Serves:
4

Preparation time:
20 minutes, plus overnight soaking

Cooking time:
2 hours 30 minutes

Prep ahead:
- Soak yellow split peas
- Cut vegetables

1 cup (200 g) dried yellow split peas, rinsed
⅓ cup (80 ml) olive oil
1.5 kg organic chicken pieces or 1 × 1.6 kg organic chicken, cut into pieces
1 eggplant, cut into large cubes
1 red onion, thickly sliced
1 teaspoon ground cumin
1 teaspoon ground turmeric (see Glossary)
1 teaspoon sweet paprika
pinch of saffron threads
6 ripe roma tomatoes, chopped
100 g pitted Sicilian green olives
100 g organic dried figs, halved
2 zucchini, cut into thick rounds
2 carrots, cut into thick rounds
1 tablespoon finely grated orange zest
2 cups (500 ml) water
3 tablespoons extra virgin olive oil
1 garlic clove, chopped
sea salt
flat-leaf parsley leaves, to serve

The addition of split peas to this gorgeous tagine makes this recipe a nutritional wonder. Using saffron is optional but it adds a beautiful depth and colour. Saffron is used in herbal medicine because it contains several plant-derived chemical compounds that are known to have antioxidant, disease-preventing and health-promoting properties.

Place the split peas in a bowl, cover with cold water and leave to soak overnight. The next morning, rinse and drain well.

Heat half the olive oil in a 5 litre flameproof casserole dish over medium heat. Working in batches, brown the chicken pieces for 5 minutes on each side, then remove and set aside. Add the remaining olive oil and fry the eggplant for 3–4 minutes until browned. Remove. Add the onion and cook for 10 minutes or until softened. Add the cumin, turmeric and paprika to the dish and cook for 2 minutes or until fragrant. Return the chicken and eggplant to the dish.

Add the saffron, tomato, olives, figs, zucchini, carrot, orange zest and water and bring to the boil, then reduce the heat and simmer for 1½–2 hours until the chicken is cooked through and tender.

While the chicken is cooking, place the split peas in a large saucepan of boiling water over high heat and cook for 40 minutes or until very soft and creamy. Drain well and transfer to a food processor, then add the extra virgin olive oil, garlic and salt to taste and process until smooth and creamy.

Serve the chicken tagine with the split pea puree, garnished with parsley.

WEEKEND PREP AHEAD #3

ROAST VEGETABLES

A batch of roast vegetables can be used in everything from salads to casserole to soups. Use any of the following:

- kent pumpkin, halved, seeded, skin left on
- carrots, halved lengthways
- zucchini, halved lengthways
- red capsicum, halved and seeded
- red onion, cut into wedges
- parsnips, halved lengthways
- small beetroot
- eggplant, cut into large bite-sized chunks
- whole small sweet potatoes
- head garlic, halved crossways
- herb sprigs (rosemary, thyme)

Preheat the oven to 200°C (400°F). Line a baking tray with baking paper. Place the veg onto the prepared tray, drizzle with olive oil and sprinkle with salt.

Bake for:

- eggplant: 30 minutes
- sweet potato: 30–40 minutes
- carrot, zucchini, capsicum, onion, parsnip, beetroot, cauliflower: 40–50 minutes
- pumpkin: 1 hour

Pack the cooled vegetables into an airtight container and store in the refrigerator for up to 5 days.

CHICKEN WITH BEETROOT & RED CABBAGE SLAW

Serves:
4

Preparation time:
25 minutes, plus
5 minutes resting

Cooking time:
10 minutes

Prep ahead:
- Halve chicken thighs
- Prepare vegetables for slaw
- Make dressing
- Roast walnuts

4 organic chicken thigh fillets, skin removed
sea salt and freshly ground black pepper
1 tablespoon olive oil
1 tablespoon ghee (see Glossary)

Slaw
2 beetroot, peeled and grated
¼ red cabbage, very thinly sliced
2 cups (80 g) shredded kale
1 red apple, cored and very thinly sliced
100 g roasted walnuts
2 tablespoons chopped flat-leaf parsley
2 tablespoons shredded mint

Dressing
2 tablespoons balsamic vinegar
2 teaspoons dijon mustard (preferably organic)
1 teaspoon poppy seeds
100 ml extra virgin olive oil
1 teaspoon raw honey (optional)
sea salt and freshly ground black pepper

I adore a good slaw – they're a great way to get more cabbage into your diet. Cabbage is a nutritional star and comes in many different varieties, each providing outstanding health benefits. Classed as a cruciferous vegetable, it is rich in vitamins C and K, folate and minerals such as potassium.

Cut each chicken thigh in half horizontally to form two thin steaks. Season well with salt and pepper.

Heat the olive oil and ghee in a large frying pan over medium heat, add the chicken and cook for 5 minutes on each side or until tender and cooked through. Remove from the pan and rest for 5 minutes.

To make the slaw, place all the ingredients in a bowl and mix to combine.

For the dressing, whisk together the vinegar, mustard, poppy seeds, oil and honey, if using. Pour over the slaw, season with salt and pepper and toss well to combine.

Slice the chicken and serve on top of the slaw.

PAN-FRIED SNAPPER WITH ZUCCHINI & PEPPERY PEA HUMMUS

Serves:
4

Preparation time:
15 minutes

Cooking time:
20 minutes

Prep ahead:
• Zest and juice lemon

4 snapper fillets, skin and bones removed
sea salt and freshly ground black pepper
3 tablespoons olive oil
2 garlic cloves, thinly sliced
3 zucchini, sliced into rounds
100 g baby spinach leaves
2 tablespoons ghee (see Glossary)
lemon slices, to serve

Peppery pea hummus
1 cup (120 g) frozen baby peas, thawed
1 cup (250 g) good-quality hummus
finely grated zest and juice of 1 lemon
1 teaspoon pink peppercorns
handful of mint leaves
sea salt

This is an ideal weeknight meal: simply make the hummus, cook the fish and veg, and dinner is served. Lovely and light, this meal still provides lasting sustenance. It's a great recipe for people who like to have a lighter meal at night, which can promote better-quality sleep for some.

To make the pea hummus, place all the ingredients in a small food processor and blend until smooth.

Pat the fish dry with paper towel and season well with salt and pepper.

Heat half the olive oil in a large deep frying pan over medium heat, add the fish and cook for 5 minutes on each side or until golden and cooked through. Remove and cover to keep warm, then wipe the pan clean with paper towel.

Heat the remaining oil in the pan over medium heat, add the garlic and cook for 1 minute or until soft. Add the zucchini and cook for 5 minutes, turning, until soft and browned. Stir in the spinach and cook until it is just wilted, then add the ghee and season with salt and pepper.

Serve the fish with the zucchini mixture and a generous spoonful of the peppery pea hummus, with lemon slices alongside.

COCONUT FISH CURRY WITH BROCCOLI

Serves:
4

Preparation time:
15 minutes

Cooking time:
35 minutes

Prep ahead:
- Prepare fish
- Cut broccoli into florets

2 tablespoons extra virgin coconut oil
½ teaspoon fenugreek seeds
1 teaspoon black mustard seeds
10 small curry leaves
1 large green chilli, halved
1 teaspoon ground turmeric (see Glossary)
1 red onion, sliced
1 × 400 ml tin coconut cream
1 tablespoon tamarind concentrate
2 ripe tomatoes, chopped
600 g firm white fish fillets, skin and bones removed, cut into large pieces
300 g broccoli, cut into florets
cooked brown rice, to serve (optional)

This fragrant Indian-inspired curry is a pretty quick recipe that doesn't compromise on flavour. I love adding turmeric to curries; it has so many health benefits, but in particular is a great anti-inflammatory aid. You can swap the fish for chicken; just pan-fry it first. Or omit the animal protein entirely and add more vegetables and some beans for a sustaining vegetarian option.

Heat the coconut oil in a large deep frying pan over medium heat, add the fenugreek and mustard seeds and cook for 3 minutes or until they begin to pop. Add the curry leaves, chilli, turmeric and onion and cook for 10 minutes or until the onion is soft.

Add the coconut cream, tamarind and tomato and bring to the boil, then reduce the heat and simmer for 10 minutes or until the tomato is soft.

Add the fish and broccoli, then cover and cook for 10 minutes or until the fish is tender.

Divide the curry evenly among four bowls (scoop out the chilli halves before serving). Accompany with brown rice, if you like.

JAPANESE EGGPLANT & SWEET POTATO WITH ZUCCHINI NOODLES

Serves:
4

Preparation time:
15 minutes

Cooking time:
30 minutes

Prep ahead:
- Cut sweet potato and eggplant
- Whisk miso mixture

300 g sweet potato, peeled and cut into chunks
500 g eggplant, cut into chunks
3 spring onions, thickly sliced
½ cup (125 ml) olive oil
sea salt
3 tablespoons shiro miso (see Glossary)
2 tablespoons apple cider vinegar
2 teaspoons tamari (see Glossary)
1 tablespoon water
100 g baby spinach leaves
4 zucchini
2 tablespoons toasted black sesame seeds
1 tablespoon freshly squeezed lemon juice
1 tablespoon ghee (see Glossary)
coriander leaves, to serve

The flavour combination of miso and eggplant is a taste sensation. Miso has a wonderful sweet/salty flavour that can be incorporated into a wide variety of recipes. As a fermented food it is particularly good for gut health. Serve this with fish, chicken or extra greens if you would like to make this more substantial – or simply enjoy as is.

Preheat the oven to 200°C (400°F) and line a baking tray with baking paper.

Place the sweet potato, eggplant and spring onion in a bowl, add the olive oil and a pinch of salt and toss to coat the vegetables. Spread them out on the prepared tray and roast for 20 minutes.

Whisk together the miso, vinegar, tamari and water, then pour over the vegetables and toss on the tray to coat them in the miso mixture. Roast for a further 10 minutes or until the miso mixture starts to caramelise.

Meanwhile, place the baby spinach in a heatproof bowl and cover with boiling water. Leave to wilt, then drain well. Using a vegetable peeler, peel the zucchini into long thin lengths, then thinly slice into spaghetti strips. Add to the spinach.

Finally, add the sesame seeds, lemon juice, ghee and roasted eggplant and sweet potato and gently toss to combine. Serve topped with coriander leaves.

BLACK SESAME SALMON ON MIXED LEAVES & AVOCADO

Serves:
4

Preparation time:
20 minutes, plus 5 minutes resting

Cooking time:
10 minutes

Prep ahead:
- Slice vegetables
- Make dressing

4 × 200 g salmon fillets, skin on
2 tablespoons black sesame seeds
2 tablespoons olive oil
120 g mixed salad leaves
6 red radishes, thinly sliced
2 tablespoons pumpkin seeds
2 tablespoons sunflower seeds
2 tablespoons linseeds
2 ripe avocados, roughly smashed
1 small fennel bulb, trimmed and thinly sliced
½ small red onion, thinly sliced
100 g cherry tomatoes, thickly sliced

Dressing
2 tablespoons freshly squeezed lime juice
1 tablespoon tamari (see Glossary)
1 tablespoon apple cider vinegar
1 tablespoon finely grated ginger
½ teaspoon sesame oil
2 tablespoons extra virgin olive oil

You can make this recipe in less time than it would take to order a takeaway. Salmon is a rich source of omega-3 fatty acids, which are essential to human health. They play a critical role in brain function, as well as growth and development, plus their anti-inflammatory action is very much needed in modern lives.

Pat the salmon fillets dry with paper towel. Scatter the sesame seeds onto a plate, then press the flesh side of each fillet into the seeds to coat.

Heat the olive oil in a large deep frying pan over medium heat, add the salmon fillets, skin-side down, and cook for 5 minutes on each side or until cooked to your liking. Remove and rest for 5 minutes while you prepare the salad.

Place the salad leaves, radish, seeds, avocado, fennel, onion and tomato in a bowl.

To make the dressing, whisk together all the ingredients. Pour over the salad and gently toss to coat.

Divide the salad among four plates, then break up the salmon into bite-sized pieces and arrange over the top.

CHAPTER FIVE

DRINKS

NOURISHING MORNING DRINK

Serves:
2

Preparation time:
5 minutes

¼ ripe avocado, sliced
1 ripe banana
2 tablespoons raw macadamias, chopped
1 teaspoon chia seeds
1 teaspoon vanilla extract
1 ½ cups (375 ml) coconut milk or milk of your choice

I'm all about variety when it comes to how we nourish ourselves and including a breakfast smoothie can offer just that. Avocados are a rich source of monounsaturated fats, which are supportive of heart health, not to mention they do wonders for our skin. This smoothie is best made just before serving as it will thicken on standing.

Place all the ingredients in a high-speed blender and blend until smooth and creamy. Pour into two tall glasses and serve.

CUCUMBER, LIME & CHIA COOLER

Serves:
4

Preparation time:
10 minutes, plus
5 minutes standing

1 tablespoon freshly squeezed lime juice
1 Lebanese cucumber, peeled
1 litre chilled water
1 tablespoon chia seeds
small handful of mint leaves, torn (optional)

The longer this drinks stands, the more gel-like the chia seeds will become. If you want a stronger mint flavour, you can blend the mint with the cucumber, but be aware that this will turn the drink from a light green to a darker green.

Place the lime juice and cucumber in a blender and blend until smooth. Transfer to a jug.

Stir in the water, chia seeds and mint, if using. Allow to stand for 5 minutes before serving.

LONG LIQUORICE MINT TEA

Serves:
4

Preparation time:
5 minutes, plus cooling time

Cooking time:
15 minutes

3 tablespoons dried liquorice root
or 3 liquorice tea bags
1 litre water
½ lemon, sliced
handful of mint leaves
200 g fresh berries of your choice
ice cubes, to serve

Liquorice is a great option for those who crave sugar as it is naturally sweet. This is the perfect afternoon drink, easy to make and equally thirst-quenching.

Place the liquorice and water in a medium saucepan. Bring to the boil, then reduce the heat and simmer for 15 minutes.

Remove from the heat, stir in the lemon, mint and berries, and set aside to cool. Serve in a jug with ice cubes.

SIX QUICK TEAS

Each recipe serves 4

IMMUNE BOOSTER

1 tablespoon grated or sliced ginger
1 tablespoon grated or sliced fresh turmeric
  (see Glossary)
zest of 1 lemon, plus lemon slices to serve (optional)
1 litre boiling water
1 teaspoon manuka honey (optional)

Place the ginger, turmeric and lemon zest in a teapot and cover with the boiling water. Leave to steep for 5 minutes. Add the honey, if using, and stir to dissolve. Strain into cups and serve with lemon slices, if you like.

DIGESTIVE TEA

1 tablespoon cumin seeds
1 tablespoon coriander seeds
1 tablespoon fennel seeds
1 litre boiling water

Place the cumin, coriander and fennel seeds in a teapot and cover with the boiling water. Leave to steep for 5 minutes, then strain into cups. Enjoy hot or cold. Best taken 20 minutes before or after meals.

BEAT THE SUGAR CRAVINGS

2 liquorice tea bags or
  1 tablespoon dried liquorice root
6 cardamom pods, bruised
2 teaspoons fennel seeds
handful of mint leaves
1 litre boiling water

Place the tea bags or liquorice root, cardamom pods and fennel seeds in a teapot and cover with the boiling water. Leave to steep for 5 minutes, then strain into cups.

>>

CALMING TEA

2 tablespoons dried chamomile flowers
4 cardamom pods, bruised
zest of 1 lemon
1 tablespoon thinly sliced fresh turmeric (see Glossary)
1 litre boiling water

Place the chamomile, cardamom pods, lemon zest and turmeric in a teapot and cover with the boiling water. Leave to steep for 5 minutes, then strain into cups.

COOLING TEA

2 lemongrass stalks, white part only, thinly sliced and bruised
grated zest of 1 lime
pinch of saffron threads
2 teaspoons fennel seeds
handful of Thai basil leaves
1 litre boiling water

Place the lemongrass, lime zest, saffron, fennel seeds and basil in a teapot and cover with the boiling water. Leave to steep for 5 minutes, then strain into cups.

SLOW-SIMMERED SPICE TEA

1 cinnamon stick
6 cardamom pods, bruised
6 cloves, bruised
1 tablespoon thinly sliced ginger
2 star anise
1. 25 litres water

Place the cinnamon stick, cardamom pods, cloves, ginger, star anise and water in a saucepan. Bring to the boil, then reduce the heat and simmer gently for 15 minutes. Strain into cups.

SPICED GINGER & LEMON NECTAR

Serves:
4

Preparation time:
10 minutes

Cooking time:
10 minutes

3 tablespoons finely grated ginger
pinch of saffron threads (optional)
1 litre water
1 tablespoon freshly squeezed lemon juice
2 teaspoons raw honey (optional)

Great to ward off colds and flu and boost the immune system, this drink features all the winter-warming favourites. Traditionally ginger has been used to ease symptoms of gastrointestinal distress and nausea. It is also believed to have anti-inflammatory properties due to its antioxidant capacity.

Place the ginger, saffron, if using, and water in a medium saucepan and bring to the boil, then reduce the heat and simmer for 10 minutes.

Remove from the heat, stir in the lemon juice and sweeten with the honey, if using.

PASSIONFRUIT LEMONGRASS MOCKITO

Serves:
4

Preparation time:
10 minutes

1 cup (250 ml) passionfruit pulp (from 11 passionfruit)
2 limes, cut into wedges
2 tablespoons finely chopped lemongrass stalks, white part only
handful of mint leaves
1 cup (250 ml) organic coconut water (see Glossary)
1 litre chilled sparkling mineral water

If you're craving an alcoholic beverage, sometimes it's actually more about quenching your thirst (although other times it's not at all – I get it!). If you can have options on hand like this refreshing and hydrating mocktail, however, at least you have the choice. Less alcohol in your life can mean better energy, sooner.

Place the passionfruit, lime, lemongrass and mint in a large bowl and use the end of a clean rolling pin to lightly crush them.

Transfer to a large serving jug. Stir in the coconut water and mineral water and serve.

WARMING GOLDEN DRINK

Serves:
4

Preparation time:
5 minutes

Cooking time:
15 minutes

Prep ahead
- Grate turmeric

2 tablespoons grated fresh red turmeric or 1 tablespoon ground turmeric (see Glossary)
1 cinnamon stick
1 litre milk of your choice
½ teaspoon ghee (optional; see Glossary)
raw honey, to taste (optional)

A wonderful bedtime drink, this can also be served as a cool drink with ice. Any leftovers can be stored in an airtight container in the refrigerator for up to 3 days and reheated.

Place the turmeric, cinnamon stick and milk in a saucepan and simmer over medium heat for 15 minutes or until the milk turns a vibrant gold. The longer you simmer the milk the stronger the flavour will be.

Remove from the heat and stir in the ghee and a little honey, if you like.

SLOW-SIMMERED CHAI

Serves:
4

Preparation time:
5 minutes

Cooking time:
1 hour 15 minutes

Prep ahead:
• Bruise spices

20 cardamom pods
20 cloves
8 black peppercorns
2 cinnamon sticks
6 star anise
1 tablespoon thinly sliced ginger
2 litres water
1 cup (250 ml) milk of your choice
raw honey, to taste (optional)
ground cinnamon, to sprinkle

You may like to add a teaspoon of ghee to your chai, if you are drinking it before bed, as this can help calm your nervous system. Coconut milk doesn't curdle when added to hot liquid and is stable at higher temperatures. Any leftover chai can be stored in an airtight container in the refrigerator for up to 3 days and reheated. This can also be served as a cool drink with ice.

Place the cardamom pods in a mortar and, using a pestle, crush them to burst the skins open and bruise the seeds. Tip into a large saucepan.

Place the cloves and peppercorns in the mortar and use the pestle to bruise them to release the essential oils. Tip into the saucepan, along with the cinnamon sticks, star anise and ginger.

Add the water and bring to the boil, then reduce the heat and simmer gently for 1 hour or until the liquid is dark brown and fragrant and has reduced by half.

Add the milk and simmer for a further 15 minutes. Remove from the heat and strain. Sweeten with honey, if using, and serve with a sprinkle of ground cinnamon and the cinnamon stick if you like.

GREEN ENERGY DRINK

Serves:
4

Preparation time:
10 minutes

3 kiwi fruit, peeled and chopped
1 avocado, chopped
small handful of flat-leaf parsley leaves
1 cup (40 g) shredded kale or baby spinach leaves
2 cups (500 ml) chilled organic coconut water (see Glossary) or water

It never ceases to amaze me the influence that real food can have on our energy. You can almost feel the nutrient boost when you drink this smoothie. It's such a great alternative to your afternoon coffee or tea. Prepare it in the morning and take it to work. Just make sure you shake it well before drinking as it will separate and settle.

Place all the ingredients in a high-speed blender and blend until smooth and creamy. Pour into glasses and serve.

ZESTY GINGER FIZZ

Serves:
4

Preparation time:
10 minutes, plus
20 minutes chilling

3 tablespoons finely grated ginger
1 tablespoon freshly squeezed lemon juice
1 litre sparkling mineral water
1 tablespoon raw honey (optional)

Essentially a more nourishing version of ginger beer, except this recipe lets the ginger do the talking (as opposed to the sweetness). As well as being great for gut health, ginger is rich in anti-inflammatory and antioxidant substances, helping you feel better from the inside out. The addition of sparkling water means you get the fizz associated with soft drinks without the preservatives and sugars.

Place the ginger, lemon juice and mineral water in a jug, add the honey, if using, and mix to combine. Chill for 20 minutes before serving.

COCO DANDY

Serves: 4

Preparation time: 5 minutes

Cooking time: 15 minutes

- 3 tablespoons dandelion tea
- 1 cinnamon stick
- 1 vanilla pod, split
- 1 litre boiling water
- 2 cups (500 ml) coconut milk
- raw honey, to taste (optional)

Dandelion tea is a fantastic alternative to coffee, and, as a bitter herb, is great for the liver. It also helps with fluid retention. You can readily find it in the tea aisle at supermarkets. While I'm not proclaiming this can replace a coffee, it certainly offers you a warming milky drink, which is often what people feel like when they order their coffee.

Place the dandelion tea, cinnamon stick, vanilla pod and water in a medium saucepan. Bring to the boil and cook over high heat for 10 minutes or until the liquid has reduced by half. Stir in the coconut milk and simmer for a further 5 minutes. Remove from the heat and stir in the honey, if using.

MMMILKY BEDTIME DRINK

Serves: 4

Preparation time: 5 minutes

Cooking time: 15 minutes

- 1 litre milk of your choice
- 1 cinnamon stick
- 1 tablespoon ground cardamom
- 1 vanilla pod, split
- 2 teaspoons ghee (see Glossary)
- 1 teaspoon pure maple syrup (optional)

A warming and nurturing bedtime drink, this is a particularly good option if you like something sweet after dinner. And as this is caffeine free, it's a great option for children as well.

Place the milk, cinnamon stick, cardamom, vanilla pod and ghee in a medium saucepan, bring to a simmer and cook gently for 15 minutes to allow the spice flavours to infuse the milk. Remove from the heat and remove the cinnamon stick and vanilla pod. Sweeten with maple syrup, if you like, and serve.

GLOSSARY

Adzuki beans (also known as aduki and azuki)

These are dark red beans widely used in Asian cooking. Adzuki beans have a wonderful creamy texture and are often used in Chinese medicine to aid digestion; some say they don't create as much gas as other beans because of this. As with some other beans, they need to be soaked overnight before cooking.

If you can't find adzuki beans you can use red kidney beans instead.

Black beluga lentils

These firm little lentils get their name because they resemble black caviar. Quick-cooking, versatile lentils that require no soaking, they retain their shape when cooked and make a wonderful addition to soups, salads and casseroles.

Available in health-food stores, you can substitute puy-style lentils if you have trouble finding them.

Buckwheat

Buckwheat is a pyramid-shaped seed that is highly nutritious and versatile. You can purchase the raw groats, which are pale green in colour (this is what I use in the recipes throughout this book). Buckwheat can be used in both sweet and savoury dishes and makes a lovely breakfast porridge.

Soaking overnight will significantly reduce cooking time; however the buckwheat will become gooey after soaking so rinse and drain it well before you cook it. Overcooking will cause the seeds to become mushy.

Buckwheat has a slightly nutty flavour and contains no gluten in spite of its misleading name. It is a great substitute for pasta or couscous. You can cook the grain ahead of time and then reheat it in a colander over a little simmering water.

Butter beans (also known as lima beans)

Butter beans are large cream-coloured kidney-shaped flat beans with a wonderful creamy texture. They need to be soaked overnight before cooking. Avoid overcooking or they will fall apart. Butter beans are delicious made into baked beans, and in casseroles, soups and salads. They take about 1 hour to cook, a little longer than most beans.

Coconut water

Coconut water is the water from the inside of a young (green) coconut. As the coconut matures the flesh absorbs the water into its flesh. Coconut water has a slightly sweet, mild flavour and is delicious in smoothies, pancakes and baking. It is relatively high in natural sugars and young coconut water will ferment quickly once opened if not stored in the refrigerator.

Dried shiitake mushrooms

Dried shiitake mushrooms are often soaked in boiling water for 20 minutes before being added to dishes, and the soaking liquid is also used to enhance the flavour of the dish.The stems are tough and chewy so remove them and discard. If you are making a soup or casserole you can break the dried mushroom caps up and add them straight to the dish. Store unused mushrooms in an airtight container in the pantry.

Ghee

Ghee is simply clarified unsalted butter. The butter has been heated to remove the milk solids and this leaves you with a clear golden liquid that has a high smoke point. Ghee is widely used in Indian cooking and it is said to improve digestion and support the immune system. It is also added to drinks to help calm the nervous system and promote sleep.

Gomasio

Gomasio is the Japanese name for a ground sesame seed and salt condiment. Very simple to make, you toast sesame seeds (black or white) and salt and then grind them to break the seeds. The ratio of sesame seeds to salt varies but usually it is about 10:1. Gomasio is great sprinkled on soups, salads or roasted vegetables.

Horseradish (fresh)

Horseradish is part of the *Brassicaceae*, or cabbage, family. The root has a peppery mustard-like flavour. To prepare fresh horseradish, remove the brown skin with a peeler and grate the flesh. Once grated, it will darken and lose flavour over time.

Hot-smoked salmon

Hot-smoked salmon has been both cooked and smoked during preparation, making it quite different from smoked salmon. It has a creamy, delicate flavour and flaky texture. Hot-smoked salmon (or trout) can be found in the refrigerated section of your supermarket.

Kaffir lime leaves

Kaffir lime leaves are dark, glossy green fragrant leaves widely used in Thai cooking. The leaves have a tough vein that should be removed before the leaves are shredded or torn and added to recipes. Leftover leaves can be frozen in airtight bags for up to 1 year. Dried kaffir lime leaves can be used as a substitute but they are nowhere near as aromatic.

Medjool dates

Medjool dates are considered to be the best of the best when it comes to dates. Larger and juicier than other varieties, medjool are favoured over other dates if you are making raw desserts, bliss balls or eating them on their own as a snack. Look for medjool dates in the fruit and vegetable department of your supermarket, not in the baking section with the other dried dates. Medjool dates are very high in natural sugars so should be eaten in moderation.

Nut butter

Nut butter can be made from a variety of nuts, including almonds, cashews, macadamias or a combination of these. Purchase good-quality nut butter from health-food stores. Store in the refrigerator as it is high in fat and will turn rancid if exposed to sunlight. Drain off excess oil or give it a good stir before using.

Puy-style lentils

Puy-style lentils are green–slate-coloured small lentils that are widely used in French cooking. They don't need to be soaked before cooking but it is a good idea to rinse them to remove any grit. Puy-style lentils only take about 15 minutes to cook – be careful not to overcook them or they will break up. They are great in salads, soups and casseroles.

Quinoa

Quinoa is a high-protein gluten-free seed that is extremely versatile and very easy to prepare. Ensure you rinse quinoa before cooking to remove the saponins, a natural insecticide that coats the seed and protects it while it is in the ground. Quinoa takes only 15 minutes to cook; it is a good idea to prepare more than you need as it is perfect the next day for salads or for a quick and easy breakfast. There are several varieties of quinoa – red, black and white – and you will also find it sold as tricolour quinoa, a combination of the three. Quinoa makes a great substitute for rice or couscous, and it can be used in both sweet and savoury recipes.

Shiro miso

Shiro miso is classified as a white miso. It is cream coloured and has a mild, slightly sweet flavour because of its limited fermentation, small amount of soy beans and large amount of rice. Shiro miso can be used in soups, marinades, casseroles, salad dressings or with vegetables. Blend miso with a little warm water to thin it before adding to soups or sauces. Store opened miso in an airtight container in the refrigerator for up to 9 months. Check the label carefully as some miso has MSG added.

Split mung dhal

Golden-coloured split mung dhal creates the creamiest of all dhal and that's why I like to use it. If you can't find split mung dhal you can use red lentils or split green mung dhal (which is the same dhal but it still has the green skins intact). Soaking the dhal in cold water for a couple of hours before cooking will help remove the unwanted skins and help it to cook faster but it isn't essential; you can simply rinse the dhal to remove any grit. Split mung dhal can be found in health-food stores. It is also known as yellow split mung dhal.

Sumac

Sumac is the ground fruit of a small shrub that is deep purple in colour. It has a slightly tangy sour citrusy flavour. Widely used in Middle Eastern recipes, it can be sprinkled over salads or used in marinades, dressings or spice mixes. You will find sumac in the spice section of your supermarket.

Tahini

Tahini is ground sesame seed paste, that comes hulled and unhulled. Hulled (with the outer skins removed) has a milder flavour and is used throughout this book. Tahini should be stored at room temperature away from direct sunlight – avoid refrigeration as it becomes solid because of its high fat content. Tahini makes a wonderful salad dressing, sweet or savoury dip and is usually thinned with water to give the desired consistency.

Tamari

Tamari is the common name given to gluten-free soy sauce. These days some brands are calling wheat-based soy sauce tamari, so make sure you check the label carefully.

Turmeric

Turmeric is a vibrant golden-coloured spice that is widely used in Indian cooking. It is growing in popularity these days as it has been recognised for its anti-inflammatory benefits. Health-food stores sell bulk ground turmeric which is often fresh and more intensely flavoured.

Fresh turmeric resembles thin ginger. It comes in two varieties, yellow and red, the red having more medicinal properties and the yellow more often used in cooking. Turmeric stains, so take care when cooking with it – I know of no miracle stain remover for it!

THANK YOU

Thank you to Ingrid Ohlsson for your vision and how you care, and to Ariane Durkin, Virginia Birch, Danielle Walker, Megan Pigott, Naomi Van Groll and the team at Pan Mac for wanting to help people cultivate better energy. I am very grateful for your support.

Thank you so much to Miriam Cannell for your wonderful edit and way with words. Thank you to Jeremy Simons, Sarah O'Brien and Tracey Pattison for a brilliant food photoshoot. You are all a joy to work with. Additional photography thanks to Rob Palmer, Michelle Noerianto and Dixie Elliott. Thank you to Emily O'Neill, Kirby Armstrong and Trisha Garner for your skilled design.

The talented and beautiful-hearted Jody Vassallo created the recipes you experience in this book. Jody and her food are gifts to this world. Thank you so much Jody.

To my team – thank you amazing humans. Particular thanks to Kate Wilkie for her brain, heart and humour and assisting so much with the creation of this book.

And to you the reader, thank you for your interest in taking even better care of yourself. May the ripple effect of your enhanced energy serve you and those around you.

ABOUT DR LIBBY

Dr Libby Weaver (PhD) graduated from the University of Newcastle in Australia with a Bachelor of Health Science Nutrition and Dietetics with Honours, and a PhD in Biochemistry after 14 years of study and clinical research. She combines this knowledge with two decades of clinical practice working with patients from all walks of life.

Her pioneering PhD research uncovered biochemical and nutritional factors that impacted children with autism and led to astounding results that forever improved the lives of those she worked with and altered the way the condition is treated in Australia and New Zealand.

A nine-time number-one bestselling author and respected international speaker, Dr Libby's expertise in nutritional biochemistry has led her to share the stage with Tony Robbins, Dr Oz, Sir Richard Branson and Marianne Williamson. In addition, she has been featured in numerous publications including *The Times*, *The Huffington Post*, *The Sydney Morning Herald* and *The Australian Women's Weekly*, and she appears regularly on breakfast television.

Her clinical experience led her to understand the intricate correlation between emotions and biochemistry, so her unique and powerful approach examines the biochemical, nutritional and emotional reasons behind what might be driving the body to behave in a certain way. She has also created a food-based nutritional supplement range called Bio Blends.

Armed with an abundance of knowledge, scientific research and a true desire to help people regain energy, vibrancy and vitality, Dr Libby empowers and inspires people to take charge of their health and happiness.

ALSO BY DR LIBBY

Accidentally Overweight
Rushing Woman's Syndrome
Exhausted to Energized
Real Food Chef
Beauty from the Inside Out
The Calorie Fallacy
Real Food Kitchen
Sweet Food Story
Women's Wellness Wisdom

INDEX

D

T

V

W

Z

First published in 2017 in Macmillan
by Pan Macmillan Australia Pty Limited
1 Market Street, Sydney, New South Wales, Australia 2000

A CIP catalogue record for this book is available from the National Library of Australia:
http://catalogue.nla.gov.au

Original design concept by Emily O'Neill
Additional cover and internal design by Kirby Armstrong
Cover concept by Trisha Garner
Photography by Jeremy Simons; additional photography by Rob Palmer
Prop and food styling by Sarah O'Brien; additional prop and food styling by Michelle Noerianto
Food preparation by Tracey Pattison; additional food preparation by Dixie Elliott
Recipe development by Jody Vassallo
Editing by Miriam Cannell and Rachel Carter
Props kindly supplied by Sarah Schembri, Chuchu and Jo Norton Ceramics
Colour + reproduction by Splitting Image
Printed and bound in China by 1010 Printing

10 9 8 7 6 5 4 3 2 1